Knowing Your Buildings: A Firefighter's Reference Guide

Knowing Your Buildings: A Firefighter's Reference Guide

CRAIG SCHWINGE

Australia • Brazil • Japan • Korea • Mexico • Singapore • Spain • United Kingdom • United States

Knowing Your Buildings: A Firefighter's Reference Guide
Craig Schwinge

Acquisitions Editor: Janet Marker
Senior Product Manager: Jennifer Starr
Editorial Assistant: Amy Wetsel
Production Manager: Mark Bernard
Marketing Manager: Deborah Yarnell
Senior Marketing Manager: Erin Coffin
Marketing Coordinator: Shanna Gibbs
Project Manager: Pre-PressPMG
Art Director: Benjamin Gleeksman
Media Editor: Tom Stover
Print Buyer: Beverly Breslin
Permissions Manager: Bob Kauser
Production Service: Pre-PressPMG
Copy Editor: Pre-PressPMG
Illustrator: Pre-PressPMG
Compositor: Pre-PressPMG

Library of Congress Control Number: 2009938523

ISBN-13: 978-1-4354-8189-3

ISBN-10: 1-4354-8189-5

Delmar
5 Maxwell Drive
Clifton Park, NY 12065–2919
USA

Printed in the United States of America
1 2 3 4 5 6 7 11 10

Dedicated
To My Family

Contents

Preface

Knowing Your Buildings: A Firefighter's Reference Guide is written for all members of the fire service to promote building literacy and to improve safety and performance on the fire ground. As a fire technology instructor in building construction at community college for the last 15 years, I realized there were few resources to help students to visualize building parts with terms that would also have an impact on their safety at structure fires. This book provides definitions on fire-service-related building construction terms with dozens of corresponding photos and drawings to:

- Provide well-defined, common terminology for typical building terms in use by construction professionals like architects, engineers, general building contractors, and code compliance building and fire inspectors.
- Provide an accepted terminology format for discussions on fire ground operations and post-incident reviews.
- Provide a visual reference guide for firefighters to reinforce building features and hazards.

For firefighters, *Knowing your Buildings* implies an active, ongoing learning process with new knowledge, construction familiarization, and observation of potentially hazardous building features. The building industry has enjoyed great economies and efficiencies in building faster, lighter, less expensive, and modular structures as a result of enhanced construction engineering and management. This is due in part to:

- New materials
- New building techniques
- New building methods
- New building designs
- New building codes including performance-based codes

Buildings have been and will always be in a dynamic process of change and renovation. Increasingly, the fire service is bringing "science to the streets," with new knowledge about burning buildings through the assistance of NIST, Homeland Security, FEMA, USFA, and NFA. It is my hope that the reader will help make this guide a "living document" by contributing additional terms and photos to improve our understanding about structures nationwide.

Some fire professionals have argued that firefighting has become more hazardous today due to light-weight engineered building features, changes in materials and methods of construction, increased fire loads in many types of structures, and less "safety margins" as a result of computer-aided design and engineering. Others have stated that "liberties" taken by contractors, inspectors, and others continue to create challenges for firefighting forces. The fire service can prevent unnecessary injuries by building upon the fire service's collective knowledge of hazardous features, how failures came to be, how to best prevent reoccurrences, and how to avoid serious mistakes by not making new mistakes.

As firefighters, we are humbled and grateful to those before us who have taught us the right thinking to make our calling safer and more effective. For this, all firefighters are indebted to those members on whose shoulders we stand. Fire service legends who have stood the test of time for uncommon sense and good thinking on the fire ground include Lloyd Layman, Emanuel Fried, Dick Sylvia, and William Clark. Other influential firefighting professionals who have shared their knowledge of building construction include Francis Brannigan, ("Undress the building," "Visualize the voids") Vincent Dunn ("Don't trust the truss," "no building is worth the life of a firefighter"), John Mittendorf ("Falling objects have the right of way"), and Michael Smith ("Awareness is not enough; we need to change our culture and approach"). The fire service thanks Alan Brunacini ("Fireground Command") and Gordon Graham ("If it is predictable, it is preventable") for their collective wisdom and sensibilities in improving fire-ground risk management. From these and many others, I have freely borrowed construction terms, insights, thinking processes, and just plain good advice and common sense about buildings that occasionally burn and fall down.

FEATURES OF THIS BOOK

There are many features incorporated into these pages that will enable you to take full advantage of the content:

- ***Construction essentials for structural firefighting success***, the opening chapter for the book, provides a foundation for the book by reviewing key points that are essential to staying safe on the fire ground. Characteristics of various types of building materials are covered, as is information on recognizing and managing risks to help firefighters understand how buildings behave under certain fire situations. An important section on lessons learned highlights actual fire injuries and fatalities drawn from NIOSH reports, and provides practical advice for predicting and more importantly, *preventing* catastrophes.
- **Current building materials and firefighting techniques** are explained and defined, to enable readers to keep pace with the latest practices in the building industry and in the fire service.
- **Extensive visuals**, including photos, graphics, and tables, illustrate various building structures and concepts to help readers make connections between what is being defined and what they will face on the fire ground.
- **Cross-references** provide a quick and easy guide to terms and concepts that relate, so that readers can more fully understand building structures.

SUPPLEMENT TO THIS BOOK

An *Instructors Resource CD* is available for instructors who wish to utilize this book within a classroom setting. This CD includes the following features:

- **Lesson Plans**, which outline various topics by subject matter and enable instructors to incorporate this book into their program
- **PowerPoint**, including full color photos along with graphics and tables, correspond to the Lesson Plans
- **Quizzes** correlate to lessons and evaluate students' knowledge of key concepts
- **Image Library**, including all photos from the book in full color, along with the graphics and tables, enables instructors to enhance their classroom presentations.

About the Author

Craig Schwinge has served twenty-eight years in the fire service and recently retired as a Captain with the San Jose (CA) Fire Department. He last worked as the Public Information Officer and developed several department documents including "Public Information and Media Relations," "Earthquake Response Plan" and "Structural Firefighting Guidelines." He also wrote *Drucker's Axioms for the Fire Service* in Fire Nuggets Fire Magazine.

He is a Certified Fire Officer with the State Fire Marshal's Office, a Hazardous Materials Specialist, Technical Rescue Technician, Emergency Medical Technician and an adjunct Instructor at Cabrillo College. He is also a licensed General Building contractor and retired Registered Environmental Assessor with the state of California. He is an alumnus of the University of California and has completed graduate work in public administration.

Acknowledgments

The publisher and author would like to thank the reviewers, who helped to shape the direction of this book:

Gary Kistner
Fire Service Management Coordinator
Southern Illinois University—Carbondale
Carbondale, IL

Ralph De LaOssa
Professor of Fire Science
Long Beach City College
Long Beach, CA

Steve Malley
Department Chair—Public Safety
Weatherford College
Weatherford, TX

Tom Sitz
Lieutenant
Painesville Township Fire Department
Kirtland, Ohio

Final acknowledgments to my wife, Patti, and daughters, Sarah and Loren, thank you for your patience and support throughout this endeavor. From here forward, I promise no more overhead questions on truss identification. To my parents, Norm and Joan, thank you for giving me every opportunity to succeed. And last to all the firefighters I have lived, worked, and learned from, and particularly to my San Jose (CA) Fire Department recruit class of June 1983 ("Learn or Burn"), thanks to help frame my thinking on what it means to endeavor to be a great firefighter. To the collective fire service of which I have been honored to be a part of, my hope is that the knowledge gained from this guide may prevent an unnecessary near miss, injury, or line-of-duty death.

Stay sharp, stay strong, stay safe.

Craig Schwinge
2009

Construction Essentials for Structural Firefighting Success

RECOGNIZING RISK

> ***"In the fields of observation, chance favors only those whose minds which are prepared."***
> ***–Louis Pasteur***

Structural firefighting has been described as both a skill and an art; it requires accurate observation, clear thinking, and appropriate decision making. Buildings are a combination of multiple variables that include size, age, condition, materials, features, engineering, and design, which creates many shades of grey that may make correct observation (*saper vedere*) difficult. Add uncontrolled fire and unknown life hazards to this view and it creates a challenging decision-making process that fire officers face daily across our nation.

The vast majority of buildings hold up to the forces of nature—wind, earthquake, fire explosion—because of engineering standards and building code enforcement. While structural failure comprises a small but significant percent of line-of-duty deaths, thousands of firefighters have been injured or exposed to near misses. As the collective fire service, we can do better. The only acceptable reason to put firefighters at a high level of risk is to effect a rescue in saving a life. Hazard avoidance and risk reduction are the hallmarks of smart firefighting. Common sense and intelligence will always be in demand at serious fires, because safety is everyone's responsibility, regardless of rank. Ignorance is costly in any endeavor, yet the costs have always been high in the fire service when things go wrong. The professional

(Courtesy of Craig Allyn Rose)

firefighter should depend upon knowledge and experience, rather than luck, for successful incident conclusion. Appropriate fire-ground risk management and decision making are a function of good judgment, experience, and training. An ongoing attitude of attention to buildings' fire hazards, and time awareness while being continuously engaged in "seeing" the entire changing scene, will reduce unforced mental errors, missteps, and unsafe operating positions—and it will reduce firefighter injuries and save lives.

Other fundamental risk-management guidelines at structure fires include:

- Recognizing hazardous building features
- Recognizing serious fire conditions (contents versus structure)
- Determining whether there is enough time and on-scene staffing to adequately perform interior attack
- Considering the stability of the structure, particularly with lightweight wood construction materials and methods found in most new residential occupancies
- Communicating the ability or inability to manage or control the interior fire environment
- Providing timely building and fire descriptions to incoming units
- Refusing to embrace the first view or snap judgment
- Ensuring situational and time awareness
- Continuing to reexamine assumptions as time and the fire progress
- Refusing to stereotype buildings, which may downplay the hazards of a particular building
- Ensuring that changes in the building or fire are properly communicated
- Reducing the number of firefighters exposed to hazardous conditions and reducing the exposure time
- Operating from neutral or safer positions whenever possible
- Operating from load-bearing positions whenever possible
- Reducing firefighter exposure to long-span areas (under, on top of, to the sides of)
- Reducing additional live and concentrated loads to weakened areas
- Reducing overloading by water to floors and flat roof areas

(Courtesy of Craig Allyn Rose)

- Refusing to allow firefighters to engage in unnecessary risky behaviors that have little benefit to final incident outcome
- Refusing to allow misunderstandings and miscommunications to continue unabated
- Refusing to focus on misleading or irrelevant information during time-critical fire-ground activities
- Refusing to base actions on the most easily remembered or recent similar incident
- Allowing for consideration of other possibilities outside one's experience or knowledge
- Reexamining assumptions, particularly if something does not fit with your thinking or actions

BUILDING BASICS

"In every art there are a few principles and many techniques." –Dale Carnegie

Most firefighters fight fires in buildings. Because there is more than one way to fight fire, much is dependent upon the specific construction features and hazards in each structure. The foundation for successful structural firefighting is built upon knowledge of simple building construction fundamentals.

For the professional firefighter, "knowing your buildings" means understanding a few basic principles:

- There is a unity that all buildings share and a set of rules that guide their use
- Buildings are designed to resist gravity, wind, earthquakes, and other **forces** 24 hours a day, seven days a week
- Buildings utilize a variety of **materials,** designs features, styles, and constructions built into a unit of interrelated and interdependent parts
- Buildings have similar features because they have similar uses
- Buildings change, as do the **fire** and **building codes** that dictate how they are built and how they are used
- The American Society for Testing and Materials (ASTM) E119 is the standard test method for fire tests of building construction and materials
- Buildings are constructed according to the code in place at the time of construction
- Many buildings are not in compliance with current building and fire codes
- "As-built" structures are the firefighter's reality, not the "as-designed" building

(Courtesy of Craig Allyn Rose)

(Courtesy of Craig Allyn Rose)

The portion of the building that distributes or carries the load is the **superstructure.**

- The superstructure is the vertical extension of a building above the **foundation**, and it has both vertical and lateral load-carrying systems
- The lateral load-carrying systems provide lateral stability to the superstructure
- Vertical load-carrying systems include **frames, bearing walls,** and combinations of both
- Horizontal load-carrying elements take vertical loads to vertical load-carrying elements and also resist lateral loads
- Horizontal load-carrying elements are normally parts of **floors** and **roofs**
- Superstructures are divided as either **wall bearing** or **frame bearing**

All buildings transmit vertical load systems to the ground through **load paths.**

- Load paths are engineered systems that allow buildings to remain upright
- Load paths utilize a "top-down" approach to viewing the **columns, beams,** floors, roofs, bearing walls, and frames and the **connections** that join them
- Fire, wind, earthquakes, and explosions may alter even properly supported load paths
- Loads impose forces within a **structural member**
- Loads and forces applied to building parts—such as walls, floors, and roofs—produce stresses
- **Tension, compression, shear,** and **bending** describe stresses both in materials and **structural elements**
- Forces are transferred from one structural element to the next
- Load paths—how forces are transmitted to the ground, and in what direction forces are transmitted—may be altered by such forces as fire, wind, earthquakes, explosions, and impacts
- Improper or overstressed load paths may be created by the forces of nature and by firefighters, which creates opportunities for failure of building elements
- Lateral load paths refer to how buildings may be "pushed over." **Shear walls, moment frames**, and diagonally braced **steel frames** resist lateral loads. Important stability factors include connections, **ductility**, and redundancy (the presence of multiple load-resisting systems)
- Moving building parts indicate load path shifts and potential failure
- Failure of a structural element changes load paths and redistributes load to areas that may not have been designed to bear such weight
- Consider load paths at all structure fires all the time
- Buildings may be altered and may not meet current code requirements

Structural elements are joined at connections:

- The most common structural elements are columns, beams, **slabs,** bearing walls and **trusses**
- Beams with greater depth can often carry greater loads
- Beams supported by columns with close **spans** can carry greater loads

- The primary structural elements are differentiated by how the load is transferred
- Other structural elements include **domes, arches, shells, cables,** and **membranes,** which support floor and roof structures
- Load paths run through connections, which can **fail** when overloaded, often before the structural element fails
- Connections are often the weak link in supporting structural elements and loads

Loads may be described by where and how they are imposed.

- Loads are classified according to how they are imposed and may be described as **live, dead, static, dynamic, impact,** or **concentrated,** among others
- Where loads are imposed is described as **axial** (through the center), **eccentric** (off center), or **torsional** (twisting)

All buildings create specific hazards to firefighting forces.

- The building creates the problem
- Firefighters engage fire in buildings of unknown burn time and often with little information about the building's contents or condition
- The arrangement, configuration, and layout within a building create specific conditions and challenges for firefighting
- The building's **contents,** structural elements, and nonstructural elements can burn and contribute to heat, smoke, and fire travel
- Contents, combustible structural elements, and building materials comprise **fire load,** which is what can burn within a structure
- **Fire** and the products of combustion obey the laws of chemistry and physics
- The forces of fire do not read blueprints or construction documents
- Fire and heat may impinge on weak links in an as-built structure
- Fire will travel in vertical and horizontal concealed spaces
- Heat transfer in buildings is primarily due to convection
- Fire will consume combustible contents and structural elements as well as heat **steel, concrete,** and **masonry** to the point of failure until the fire is cooled, smothered, or runs out of fuel
- Structural elements are parts of the superstructure, which may be combustible
- Nonstructural elements include finishes, **ceilings, HVAC systems,** and interior arrangements such as window and door openings, hallways, stairwells, and shafts, which provide horizontal

(Courtesy of Craig Allyn Rose)

(Courtesy of Craig Allyn Rose)

and vertical avenues for fire—and the toxic by-products of combustion—to travel

- The way a building is constructed, as well as the fire load's rate of heat release, will affect the extent and severity of fire growth
- **Voids** conceal building components
- Fire travel in voids and **combustible concealed spaces** has been the cause of many large-loss fires
- **Penetrations** in **compartmentalized attics** allow heat, smoke, and fire to spread
- Structural integrity is more often compromised by alterations and renovations than by exceeding the time of fire endurance of structural elements
- Under fire conditions, combustible structural components cannot be assumed to retain structural stability
- Fire damage to structural members equals the loss of the load-carrying capability, which equals a change in load paths and forces that may result in failure
- Failure of a structural element changes load paths and redistributes loads to areas that may not have been designed to carry such weight
- Long-span structural elements that fail create large failure areas

Firefighting forces deal with many hazardous building features.

- **Unreinforced masonry walls**
- **Parapet walls**
- **Facades**, overhangs, and **marquees**
- Voids and concealed spaces
- Unprotected steel and steel bar joists
- Unprotected floor and roof support systems, particularly lightweight wood trusses and I-beams
- Large buildings without sprinkler systems
- Long-span floor and roof support systems
- Noncompartmentalized, large-area attics
- Large, open interior areas and hallways
- Unprotected vertical shafts
- Window and door locations that restrict entry or egress and/or allow for fire, heat, and smoke extension
- Combustible, non–fire stopped load-bearing elements

In addition to having hazardous features, buildings may also harbor hazardous conditions:

- Poor condition due to lack of maintenance
- Deficiencies in construction

- Damage or deterioration to the superstructure
- Distorted structural members (sagging, bowing, leaning)
- Out-of-plumb columns and walls
- Out-of-level floor support systems
- Movement of stationary building components
- Rapid change in heat, smoke, or fire conditions during fire-ground operations

MATERIAL BASICS

Wood Basics

The primary materials used in structural elements are wood, steel, masonry, and concrete. Wood has certain qualities that make it an ideal building material.

- Inexpensive, strong, easy to work with, long-lasting if properly protected and maintained, but combustible
- Framing uses softwood like Douglas fir, cedar, hemlock, spruce, and redwood; Douglas fir is popular due to its strength
- **Lumber** may be classified as **timber, dimension**, and **boards**
- Timber is lumber that is 5 inches or thicker
- Dimension is 2 to 4 inches thick and of any width
- Boards are 1 to 1.5 inches thick and 2 inches or wider
- A visual grading system is based on size and use
- Moisture and shrinkage in unseasoned horizontal members can make floors uneven
- "Green" structural members can also warp, twist, and shrink, which may cause connectors to fail
- With wood beams, the size of the beam needed to span any opening depends on the load it has to carry and the strength of the lumber. Allowable stresses and capacities for specific wood types and structural members are specified in engineering documents and reference texts
- For wood posts, the most common size for residential and commercial buildings are 4 × 4 feet and 4 × 6 feet
- Wood **sheathing** means plywood and oriented strand board (OSB) that replaces boards for most applications. Sheathing is normally sold in 4 × 8 sheets of varying thickness, with three to five layers and glue between layers

Steel Basics

- Versatile, uniform quality, great strength in compression and tension
- Carbon steel, ASTM grade A-36, is the standard grade used in building construction
- Steel loses strength at 1000 degrees Fahrenheit (F)
- It is most popular in commercial and industrial buildings
- Steel is increasingly used for beams and columns in wood-frame residences two stories or higher to span large openings (windows and garage doors) and carry heavier loads
- Used more in multistory complexes and "soft-story" townhouses and apartments
- To protect against fire, encasement—layers of **plaster**, concrete, or **gypsum board**—or **sprayed-on fire-resistance rating** is achieved by the thickness of the cover
- Many buildings four stories or taller are built using steel columns, beams, and **decking**

(Courtesy of Craig Allyn Rose)

(Courtesy of Craig Allyn Rose)

- Decking supports concrete slabs and **suspended ceilings**
- Exterior walls may be **metal panels,** precast **curtain walls,** or masonry
- Common hot-rolled steel shapes are wide-flange beams (W) and American Standard Beam (S) for columns and beams typically 4" to 36" deep
- Cold-formed structural shapes—**purlins, girders, struts,** and diagonal bracing—are used for secondary members in most pre-engineered metal buildings
- Steel studs are used extensively in Type I Fire Resistive and Type II Non-combustible buildings as well as for other building types, particularly in tenant improvement (T&I) for commercial occupancies
- Size and gauge of steel studs depends upon the location and loading conditions
- Steel columns typically come in three shapes: wide flange, pipe, and structural tubing and are usually fabricated with a base plate and a connection at the top to support a beam
- **Steel decking** is typically sold as flat or ribbed 12- to 16-gauge sheets and used for floors and roofs with poured-over concrete slabs

Masonry Basics

- Durable, with fire- and heat-resistant and sound-insulating properties
- Fast and easy to put up with little maintenance
- Choices of texture, color, style, and pattern
- Heavy material, high compression strength, requires **steel reinforcement** in earthquake zones
- Little tensile or flexural strength
- Typical examples include **concrete block,** brick, and stone with **mortar** and/or **grout**
- Used in load-bearing and nonbearing walls for interior and exterior applications
- May be used below and above grade for **piers** and columns, **fire walls,** and curtain walls
- ASTM creates grades to describe structural properties of masonry, sand, **cement,** lime, and **rebar**
- **Concrete masonry units (CMUs)** are typically 8 × 8 × 16 feet and are used for wall thicknesses of 8 to 16 inches
- Steel **connectors** are used to join masonry walls with wood and steel roofs and floors
- Unprotected wide-flange beams used to support floors and roofs are susceptible to elongation when heated and may cause failure of the masonry wall

(Courtesy of Craig Allyn Rose)

Concrete Basics

- **Noncombustible**, heavy, and brittle material with great compressive strength
- Uses include floors, walls, roofs, columns, and beams
- Steel reinforcement (rebar, wire mesh) is necessary for floors, **foundations,** columns, and beams to create tensile strength
- **Prestressing** develops greater load-carrying capabilities with less weight
- **Pretensioning** is used for **precast concrete**
- **Posttensioning** is used for **cast-in-place concrete**
- The thickness of exterior concrete walls depends upon the design load and **fire-resistance** requirements
- A 3.5-inch concrete wall provides a 1-hour rating
- A 5-inch concrete wall provides a 2-hour rating
- A 6-inch concrete wall provides a 3-hour rating
- A 6.5-inch concrete wall provides a 4-hour rating
- Precast **tilt-up walls** are usually cast off-site at a manufacturing plant and are erected, or tilted-up, into place with a crane
- Walls are cast on-site, using the building floor slab as the casting platform, and tilted up with a crane
- **Wall panels** are lifted and then braced to the floor slab
- Panels are joined together by welding them to steel columns or joining them to cast-in-place concrete **pilasters**
- **Concrete beams** are typically rectangular and are used to carry floor and roof loads
- Concrete beams are reinforced with steel to resist **tension** (the stretching of the lower part of the beam) and at the top of the beam near the supports, the **reaction points**
- Prestressing beams is an engineering technique that creates greater load-carrying capacity
- Many parking structures have pretensioned and posttensioned structural elements

LESSONS LEARNED

"Fools learn from experience, wise men learn from the experience of others" –Otto von Bismarck

The National Institute for Occupational Safety and Health (NIOSH) is a federal agency tasked to write reports after worker injuries and/or deaths, including those of firefighters. The National Institute of Standards and Technology (NIST) does fire dynamics testing and assists NIOSH with research and investigation. Common issues revolving around firefighter line-of-duty deaths and near misses include stress-related cardiac events; lack of fire-ground accountability;

(Courtesy of Craig Allyn Rose)

becoming lost, separated, or running out of air; hostile fire conditions; and structural failure.

The following are recommendations shown on firefighter fatality reports from 1996 through 2008. Note that NIOSH cites several that are specific to the failure of floors, walls, and roofs at structure fires:

- Train all firefighting personnel in risks and hazards related to structural collapse
- Recognize potential collapse hazards, which include parapet walls, state of disrepair, and full fire involvement
- Establish and monitor a clearly marked collapse zone for structures that have become unstable due to fire damage
- When analyzing for potential building collapses, recognize structural defects and large body of fire in old structures
- Pay attention to non–automatic-sprinklered commercial retail stores
- Modifications/renovations/changes in occupancy without the benefit of review or inspection by building professionals may decrease the structural integrity of supporting members

F2008-21: "Volunteer Fire Chief Killed when Buried by Brick Parapet Wall Collapse – Texas," *http://www.cdc.gov/niosh/fire/reports/face200821.html*

F2007-08: "Career Firefighter Dies when Trapped by Collapsed Canopy during a Two-Alarm Attached Garage Fire – Pennsylvania," *http://www.cdc.gov/niosh/fire/reports/face200708.html*

F2007-01: "Career Firefighter Dies and Chief Is Injured when Struck by 130-Foot Awning that Collapses during a Commercial Building Fire – Texas," *http://www.cdc.gov/niosh/fire/reports/face200701.html*

F2006-27: "Floor Collapse at Commercial Structure Fire Claims the Lives of One Career Lieutenant and One Career Firefighter – New York," *http://www.cdc.gov/niosh/fire/reports/face200627.html*

F2006-26: "Career Engineer Dies and Firefighter Injured after Falling through Floor while Conducting a Primary Search at a Residential Structure Fire – Wisconsin," *http://www.cdc.gov/niosh/fire/reports/face200626.html*

F2006-24: "Volunteer Deputy Fire Chief Dies after Falling through Floor Hole in Residential Structure during Fire Attack—Indiana," *http://www.cdc.gov/niosh/fire/reports/face200624.html*

F2006-07: "Two Volunteer Firefighters Die when Struck by Exterior Wall Collapse at a Commercial Building Fire Overhaul – Alabama," *http://www.cdc.gov/niosh/fire/reports/face200607.html*

F2004-37: "Volunteer Chief Dies and Two Firefighters Are Injured by a Collapsing Church

Facade – Tennessee," *http://www.cdc.gov/niosh/fire/reports/face200437.html*

F2003-18: "Partial Roof Collapse in Commercial Structure Fire Claims the Lives of Two Career Firefighters – Tennessee," *http://www.cdc.gov/niosh/fire/reports/face200318.html*

F2002-50: "Structural Collapse at an Auto Parts Store Fire Claims the Lives of One Career Lieutenant and Two Volunteer Firefighters – Oregon," *http://www.cdc.gov/niosh/fire/reports/face200250.html*

F2002-44: "Parapet Wall Collapse at Auto Body Shop Claims Life of Career Captain and Injures Career Lieutenant and Emergency Medical Technician – Indiana," *http://www.cdc.gov/niosh/fire/reports/face200244.html*

F2002-40: "Career Firefighter Dies after Roof Collapse following Roof Ventilation – Iowa," *http://www.cdc.gov/niosh/fire/reports/face200240.html*

F2002-32: "Structural Collapse at Residential Fire Claims Lives of Two Volunteer Fire Chiefs and One Career Firefighter – New Jersey," *http://www.cdc.gov/niosh/fire/reports/face200232.html*

F2002-11: "One Career Firefighter Dies and a Captain Is Hospitalized after Floor Collapses in Residential Fire – North Carolina," *http://www.cdc.gov/niosh/fire/reports/face200211.html*

F2002-07: "One Career Firefighter Dies and Another Is Injured after Partial Structural Collapse – Texas," *http://www.cdc.gov/niosh/fire/reports/face200207.html*

F2002-06: "First-Floor Collapse during Residential Basement Fire Claims the Life of Two Firefighters (Career and Volunteer) and Injures a Career Firefighter Captain – New York," *http://www.cdc.gov/niosh/fire/reports/face200206.html*

F2001-16: "Career Firefighter Dies after Falling through the Floor Fighting a Structure Fire at a Local Residence – Ohio," *http://www.cdc.gov/niosh/fire/reports/face200116.html*

F2001-09: "Volunteer Firefighter Dies and Another Firefighter is Injured during Wall Collapse at Fire at Local Business – Wisconsin," *http://www.cdc.gov/niosh/fire/reports/face200109.html*

F2001-03: "Roof Collapse Injures Four Career Firefighters at a Church Fire – Arkansas," *http://www.cdc.gov/niosh/fire/reports/face200103.html*

99-F04: "Roof Collapse in Arson Church Fire Claims the Life of Volunteer Firefighter – Georgia," *http://www.cdc.gov/niosh/fire/reports/face9904.html*

99-*F03*: "Floor Collapse Claims the Life of One Firefighter and Injures Two – California," *http://www.cdc.gov/niosh/fire/reports/face9903.html*

98-F20: "Firefighter Dies while Fighting Warehouse Fire when Parapet Wall Collapses – Vermont," *http://www.cdc.gov/niosh/fire/reports/face9820.html*

98-F17: "Sudden Floor Collapse Claims the Lives of Two Firefighters and Four Are Hospitalized with Serious Burns in a Five-Alarm Fire – New York," *http://www.cdc.gov/niosh/fire/reports/face9817.html*

97-F04: "Floor Collapse in a Single-Family Dwelling Fire Claims the Life of One Firefighter and Injures Another – Kentucky," *http://www.cdc.gov/niosh/fire/reports/face9704.html*

96-F17: "Sudden Roof Collapse of a Burning Auto Parts Store Claims the Lives of Two Firefighters – Virginia," *http://www.cdc.gov/niosh/fire/reports/face9617.html*

MANAGING THE RISK

> ***"Things that go wrong are predictable, and predictable is preventable." –Gordon Graham***

To build the foundation for fire-ground success, departments need to deliver effective firefighters capable of recognizing hazardous building features. Predicting adverse outcomes is less precise. Deficiencies in construction, poor building maintenance, a fire's long-burn time due to combustible bearing elements, extension into concealed spaces, and exposure to unprotected, lightweight structural members have all been factors that have led to "bad building behaviors."

Firefighting risk management is hard to achieve, because the job has a risk profile of few options within a compressed time frame, a limited upside, and an unlimited downside potential. The downside risk is always severe and must be avoided, unless the actions are in the course of saving lives. "Good" fire officers are like hockey or soccer goalies: always in the game and frequently positively at the heart of it. Knowing what is accurate about the burning building, and knowing what you don't know about that specific burning building, is a rational approach to answering the question: "What are the consequences if I am wrong?"

The fire service is growing to understand that "acceptable losses" for buildings are to be expected; acceptable losses for firefighting forces are never an acceptable loss. As human beings, we make mistakes;

as firefighters, we need to avoid the serious ones. The fire-service community will be less willing to accept risky behaviors and "cultural norms" by those engaged in non-rescue firefights with a low likelihood of a favorable outcome.

Predictions of structural failure due to fire include the following:

- Buildings with hazardous features will continue to burn and fail, which will result in injury or near misses to firefighters
- Residential occupancies will continue to have the greatest number of fires and the greatest number of fire fatalities
- Large-loss fires will continue to occur in buildings that lack automatic sprinkler systems and compartmentation in large-volume attic areas
- Unreinforced masonry buildings (URMs) will continue to be one of the most hazardous buildings for firefighting forces
- The building industry will continue to design and build lightweight engineered structures and elements for greater economy and speed of construction
- Engineered structures will have more "disposable" structural elements after fire impingement or failure
- Incidents of failure in Type III and Type V buildings with lightweight wood trusses and I-beams will increase
- Some arriving officers will not provide an adequate size-up or a good report on conditions, which can begin a compounding chain of errors that result in poor fire-ground risk management
- Some firefighters will continue to make misjudgments about burning buildings and time monitoring and thus be exposed to near misses
- Adverse outcomes are to be expected at structure fires with hazardous building features combined with poor fire-ground risk management and decision making

Preventions for line-of-duty deaths and near misses due to structure failure must include the following:

- Prevention of fire and providing life safety education and training for building occupants remains a priority
- Enforce code requirements for automatic sprinkler and detection systems through regular inspections
- Propose local ordinances for residential sprinklers and to retrofit large buildings with automatic sprinkler systems

(Courtesy of Craig Allyn Rose)

- Modify codes to require 1-hour fire-rating protection for all lightweight wood floor and roof systems
- Compartmentalize large concealed spaces by eliminating draft-stop exemptions and require compartmentalization for large area concealed attic spaces and modify codes to increase draft stopping requirements
- Require prefire analysis documentation as part of the building permit process
- Develop department-specific structural firefighting policies and guidelines appropriate to the risk hazards and staffing levels
- Train company officers and incident commanders on fire-ground risk management, decision making, and supervision
- Provide timely discipline for infractions that needlessly endanger firefighters' lives
- Provide all fire personnel with ongoing, realistic training that includes simulation assessments and that focuses on identification of hazardous building features in various building types
- Educate firefighters to become more "building literate" by increasing building familiarization and by performing walk-throughs and preplanning
- Utilize the knowledge and technical skills of local architects, structural engineers, and general contractors to improve safety and performance at structure fires
- Ensure that standard operating procedures match resource availability and capability for various buildings
- Perform critical task analysis for different buildings to determine capabilities and expectations for fire control
- Anticipate increased levels of acceptable losses to buildings and contents as a result of increased fire loads and engineered structures
- Anticipate more defensive operation firefights
- Ensure that fire companies operate within predetermined risk-management guidelines
- Ensure that fire companies do not operate outside safety boundaries for non–life saving incidents
- Anticipate extension of fire into concealed spaces or voids, particularly into large, undivided attic spaces
- Anticipate failure of walls when unprotected steel structural members are exposed to fire
- Anticipate failure of unprotected steel bar joists when impinged upon by fire
- Anticipate failure of unreinforced masonry walls (URMs) when floor and roof systems fail
- Anticipate deficiencies and alterations in buildings that are in poor condition and that may create failure points, which are weak links
- Anticipate earlier roof failure in structures with lightweight, combustible, prefabricated trusses and I-beams
- Anticipate more "disposable" buildings and elements in areas where performance standard codes are in place
- Ensure that appropriate collapse zones and/or hazardous feature areas are established
- Reduce oversights and mental errors by estimating fire spread and extension, monitoring time, and identifying building and fire hazards; an unabated chain of errors can be avoided by providing clear and comprehensive size-ups by first arriving companies that will provide different time and location perspectives and may identify additional hazardous building features and fire conditions
- Put in place control measures and mitigators to limit firefighter exposure to unnecessary hazards, and request regular updates from officers in, on, and around the fire building
- During rescue operations, limit exposure by putting the least number of firefighters at high levels of risk for the shortest amount of time possible, and prevent poor rescue outcomes by initiating search in areas of the building with the highest probability of live victims, anticipating time and air awareness of working companies, supporting the rescue operation with necessary resources and means, protecting avenues of egress, providing timely ventilation, establishing a two-in–two-out rule and/or a rapid intervention company, and securing avenues of escape by laddering, creating doors out of windows, and other appropriate actions

Abatement a lessening in degree or intensity

Accessible describes a site, building, facility, or other structure that can be approached, entered, and used by persons with physical disabilities

ACI American Concrete Institute

Acoustic Ceiling Tiles low-density lightweight tiles designed to absorb sound; typical sizes range from 1/2 to 3/4 inches thick and from 12 to 60 inches across; tiles may be set with adhesives, or they may be fastened to furring or suspended from above to provide a concealed space for mechanical ductwork, electrical conduit, and plumbing lines; light fixtures, sprinkler heads, fire-detection devices, and sound systems can be recessed into the ceiling plane, and these tiles can be fire-rated and may provide fire protection for the supporting floor and roof structure (see also *T-Bar Ceiling*)

Acoustic Ceilings provide sound treatment as well as a finished ceiling surface; may be set with adhesives, sprayed on, nailed to furring, or suspended (T-bar ceilings)

Acoustic Tile square or oblong sheets of fibrous material designed to absorb sound in various sizes

Admixture a material other than water, aggregate, or hydraulic cement, used as an ingredient in concrete to modify its properties or those of the hardened product; typical admixtures include air-entraining agents that increase workability and produce lightweight, insulating concrete; may also include accelerators and retardants, surfactants to reduce the surface tension of the mixing water, coloring agents, and water-reducing agents to increase strength

Aggregate various granular materials such as sand, crushed stone, and gravel that is added to cement to make concrete

Agricultural Building designed and constructed to house farm implements, hay, grain, poultry, livestock, or other horticultural products; not to be used by the public or for human habitation

Air Shaft a vertical opening penetrating the floors of a building, designed to provide air and light to window openings in the walls of the shaft; typical for older, multiresidence buildings in urban settings

AISC American Institute of Steel Construction

Aisle an exit-access component that defines and provides a path of egress

Alley any public thoroughfare greater than 10 feet but less than 16 feet in width that has been deeded for public use

⚠ **Alteration** any change, addition, or modification in construction or occupancy

Anchor a metal device used to hold down the ends of trusses or heavy timber members at the walls; anchor bolts are used in foundations to secure the mud sill to the top of the foundation

Anchorage to tie roofs to walls and columns and to tie walls and columns to foundations **(Figure A-1)**

Annunciator a unit containing one or more indicator lamps, alphanumeric displays in which each indication provides status information about a circuit condition or location

ANSI American National Standards Institute

APA Performance-Rated Panels plywood manufactured to the structural specifications and standards of the Engineered Wood Association (APA)

Arch a typically curved structural member spanning an opening and serving as a support for the wall or other weight above the opening. Developed by the Romans, an arch combines the function of a beam and a column for longer spans between columns. Arches produce an outward thrust, as well as a downward force, at the supported ends, requiring the base to be braced or tied, often with abutments or buttresses. Rigid arches are curved, rigid structures of timber, steel, or reinforced concrete capable of carrying some bending stresses. Masonry arches are constructed to individual wedge-shaped stones or bricks. The removal or destruction of any part of an arch may cause the entire arch to collapse **(Figure A-2)**

Architect a person who designs and supervises the construction of buildings or other large structures

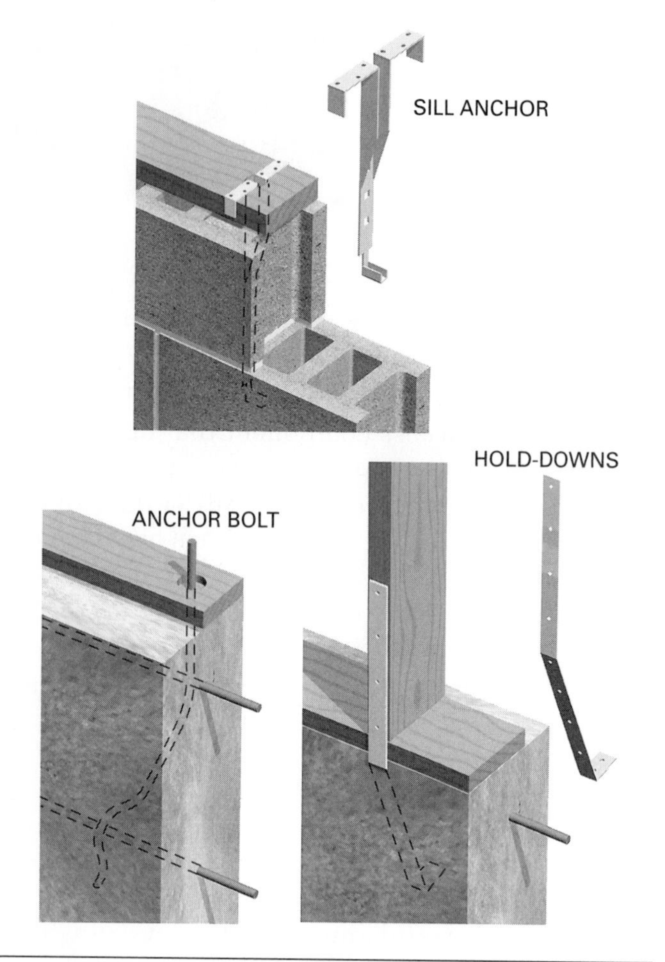

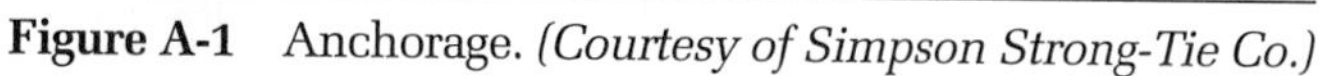

Figure A-1 Anchorage. *(Courtesy of Simpson Strong-Tie Co.)*

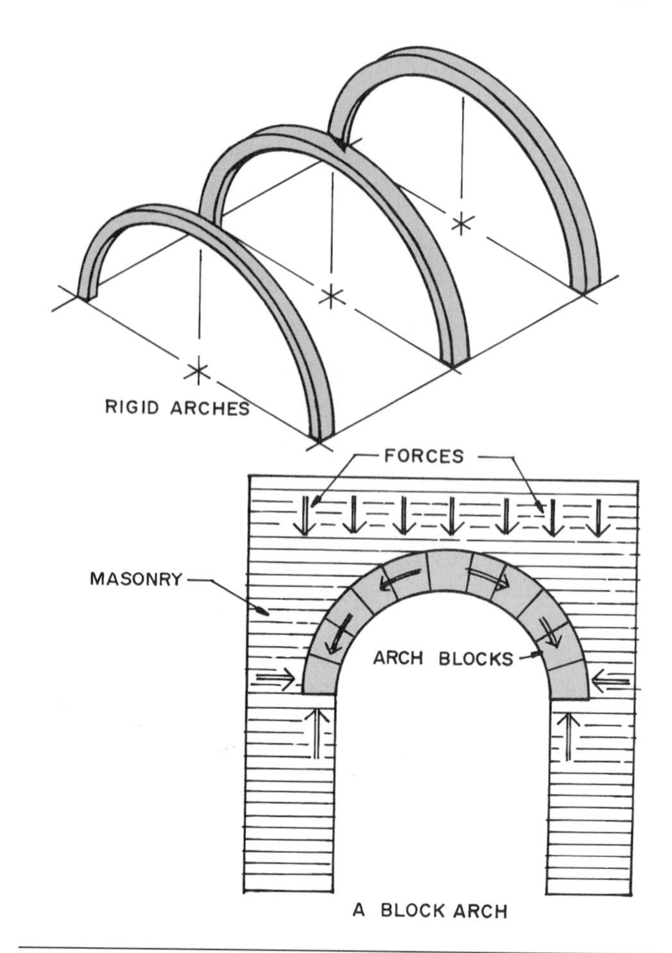

Figure A-2 Arches May Be Rigid or Block Types.

Anchorage to tie roofs to walls and columns and to tie walls and columns to foundations **(Figure A-2)**

Area of Refuge an area with direct access to an exit or an elevator, where persons unable to use stairs can stay temporarily in safety to await instructions or assistance during an emergency evacuation

Armored Cable metal conduit used to protect electrical wiring

Asbestos a noncombustible, fibrous material once commonly used in floor tiles, as a wrap for boiler pipes, and as a fire retardant material; asbestos is infrequently used today due to its cancer-causing characteristics

Asbestos-Cement Roofing Shingle a prepared roof material that combines asbestos and cement to create a noncombustible covering

Asphalt a tarlike substance consisting chiefly of hydrocarbons, used for pavements and as waterproof cement (see also *Bituminous*)

Asphalt Shingle/Composition Shingle a shingle made from asphalt and structural fibers that is bonded together and covered with a protective granular coating; used on roof slopes of at least two units vertical in 12 units horizontal and fastened to solidly sheathed decks; also known as prepared roofing (see also *Fiberglass Shingles*)

ASTM American Society for Testing and Materials; ASTM E119 is the standard test method for fire tests of building construction and materials

Atruim a vertical opening through two or more stories that was originally created for light and ventilation, but is now used more frequently for architectural effect; as defined in NFPA 92B, Guide for Smoke Management Systems in Malls, Atria, and Large Areas, this large, uncompartmented space can allow smoke to move and accumulate without restriction

Attic the space between the uppermost ceiling and the roof of a building

Attic Ventilation a design feature that provides air exchange in concealed roof spaces via eave and ridge vents; eave or soffit vents may consist of a continuous screened vent slot or a metal vent strip, installed in

Figure A-3 Attic Ventilation.

the eave soffit, or a series of circular plug vents in frieze boards; ridge ventilation may be provided by a continuous ridge vent or by louvers in the gable-end walls of unheated attics **(Figure A-3)**

Automatic Sprinklers varying types of piping systems that carry water to heads which allow water to come out and extinguish fires; the system is usually activated by heat from a fire and discharges water over the fire area

Automatic Venting a roof opening triggered by heat or smoke

Awning a shelter supported entirely from the exterior wall of a building

Axial Load a load passing through the center of the mass of the supporting element, perpendicular to its cross section

B

Backing the surface or assembly to which veneer is attached

Balcony a projecting platform, often cantilevered

⚠ **Balloon Framing** a wood framing method in which studs raise the full height of the frame from the sill plate to the roof plate, with joists nailed to the studs and supported by sills or ribbons let into the studs; balloon framing is rarely used today, as limited fire stops allow unobstructed vertical fire travel; lack of fire-stopping on floor plates allows heat, smoke, and fire run of the entire wall and into the attic **(Figure B-1)**

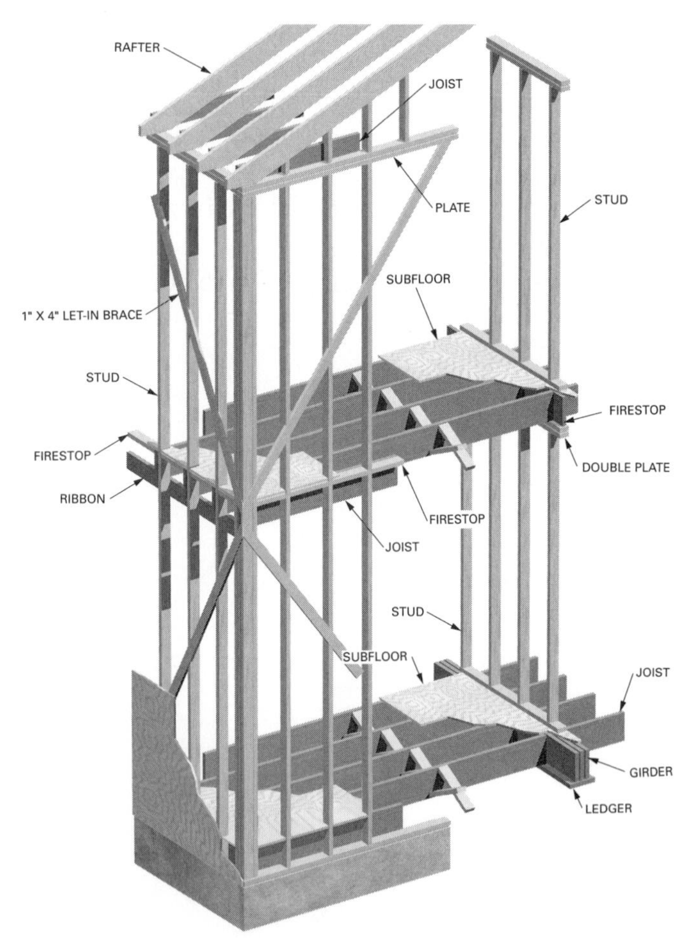

Figure B-1 Balloon Framing.

Bar Joist a steel joist with bars in the vertical web space; a truss similar to a parallel chord truss **(Figure B-2)**

Figure B-2 Bar Joist.

Barge Rafters / Fly Rafters the end rafters in the part of a gabled roof that projects beyond the gable wall; lookouts connect these rafters to double common rafters, which create rake overhangs

Baseboard typically a wood trim that conceals and finishes the joint where sidewalls meet the floor

Basement any floor level below the first story in a building

Bay the space between beams or rows of columns

Bay Window a window assembly whose maximum horizontal projection is no more than 2 feet from the plane of an exterior wall, elevated above the floor level, or a window supported on a foundation extending beyond the main walls of a building

Beams rigid structural members subjected to loads perpendicular to their length; beams transmit such forces to a point of support. The primary design characteristic of beams is their ability to resist deflection, bending and resisting moment as well as

B

their capacity to withstand shear and stresses; the load-carrying capacity of beams is influenced by dimension, span, design, and loading; the greater the depth of the beam, the greater the load-carrying ability or ability to span greater distances; beam materials include steel, wood and laminated wood, and reinforced concrete; beam support methods include simple, cantilever, suspended, fixed-end, and continuous. Simple beams rest on supports at both ends with ends free to rotate; a cantilever is a projecting beam supported at only one fixed end; fixed-end beams have both ends restrained; continuous beams extend over more than two supports, typically supported at both ends and at the center **(Figure B-3)** (see also *Girders, Joists, Rafters, Headers, and Lintels*)

Beam Forces the "pushing out" force beams exert on their ends **(Figure B-4)**

Beam Pocket a space in walls to support the ends of wood beams or joists; a space typically thought of in masonry or concrete walls that may support restrained or unrestrained beams **(Figure B-5)**

Bearing a point, surface, or mass that supports weight, typically viewed as the contact area between a beam or a truss and a wall, column, or other underlying support

Bearing Plate a steel plate placed under a beam, column, or truss to distribute the end reaction from the beam to the supporting member

Figure B-3 Beams.

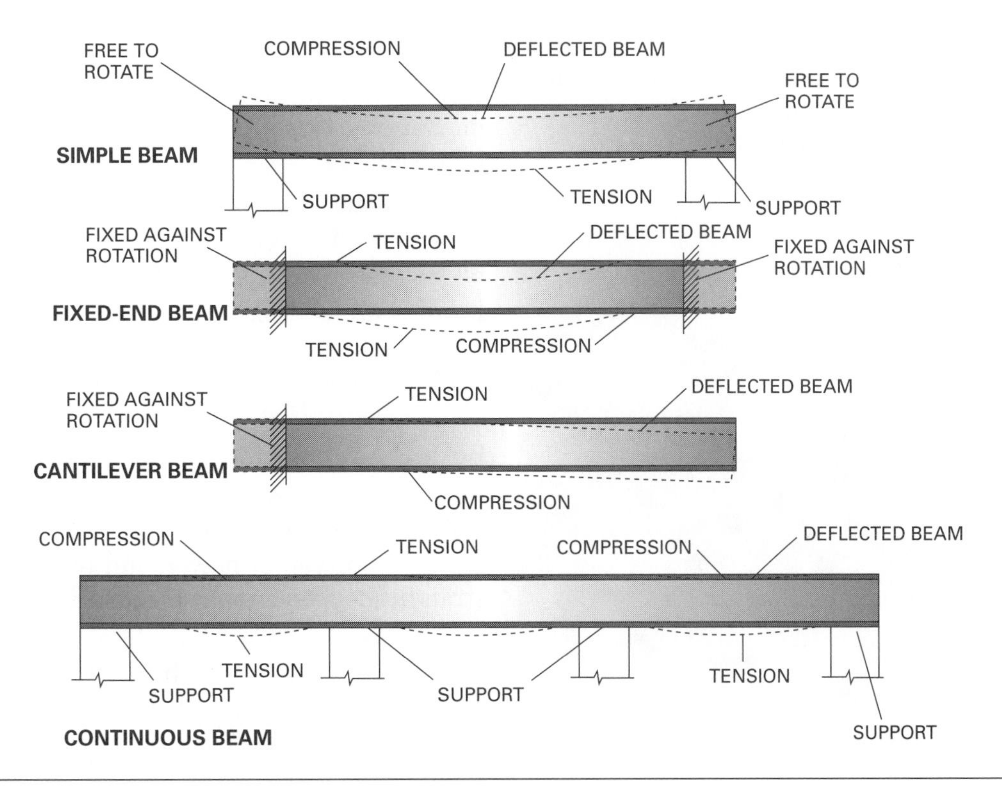

Figure B-4 Beam Forces. *(Courtesy of* Basic Engineering for Builders, *by* Max Schwartz, *published by Craftsman Book Company.)*

B

Figure B-5 Beam Pocket.

Bearing Wall a wall that supports all, or a portion, of a suspended load such as a floor, roof, beam, truss, or other wall; may be referred to as a *load-bearing wall* or *structural wall* **(Figure B-6)** (see also *Load-Bearing Wall*)

Figure B-6 Bearing Wall.

Bearing Wall System a structural system with bearing walls that provides support for all or major portions of the vertical loads; shear walls or braced frames provide seismic force resistance (see also *Shear Walls* and *Braced Frames*)

Bending Moment the load that will bend or break the beam

Bent braced or rigid frames designed to carry vertical and lateral loads that transverse the length of a framed structure; may also be a line of columns going in any direction

Bird's Mouth a right-angled notch cut on the underside of a rafter to fit over a beam or over the top plate of a stud wall frame **(Figure B-7)**

Bird Stop/Frieze Board typically a wooden member that closes the gap between rafters or trusses on the outside bearing wall

Bitumen an asphalt compound used to improve a material's water resistance and toughness

Bituminous of, or pertaining to, asphalt (see also *Asphalt*)

Blocking a number of small wood pieces between studs and joists to space, join, or reinforce structural members of a building frame; blocking provides a nailing surface for finish materials and prevents the passage of fire between the members

Board lumber less than 2 inches thick and at least 2 inches wide; boards are graded for appearance, rather than strength, and are used as siding, subflooring, and interior trim

Bond the adhesion between masonry units and mortar or grout, also the systematic lapping of brick in a wall

Bond Beam a horizontal grouted element in masonry in which reinforcement is embedded, typically used for foundations and to span openings in walls

Bond Course a brick layer that is laid perpendicular to the running course; it signifies that the brick wall is a load-bearing wall as compared to a brick fascia wall (see also *Header Course*)

Bottom Chord the lower horizontal member of a truss

⚠ **Bowstring Truss** a truss whose upper chords are curved or bowed and whose lower chord is straight (see also *Truss* **(Figure T-7A)**)

Box Beam a beam having a hollow, rectangular cross section made by gluing two or more plywood or oriented strand board webs to sawn or laminated veneer lumber flanges; box beams are engineered to span up to 90 feet

B

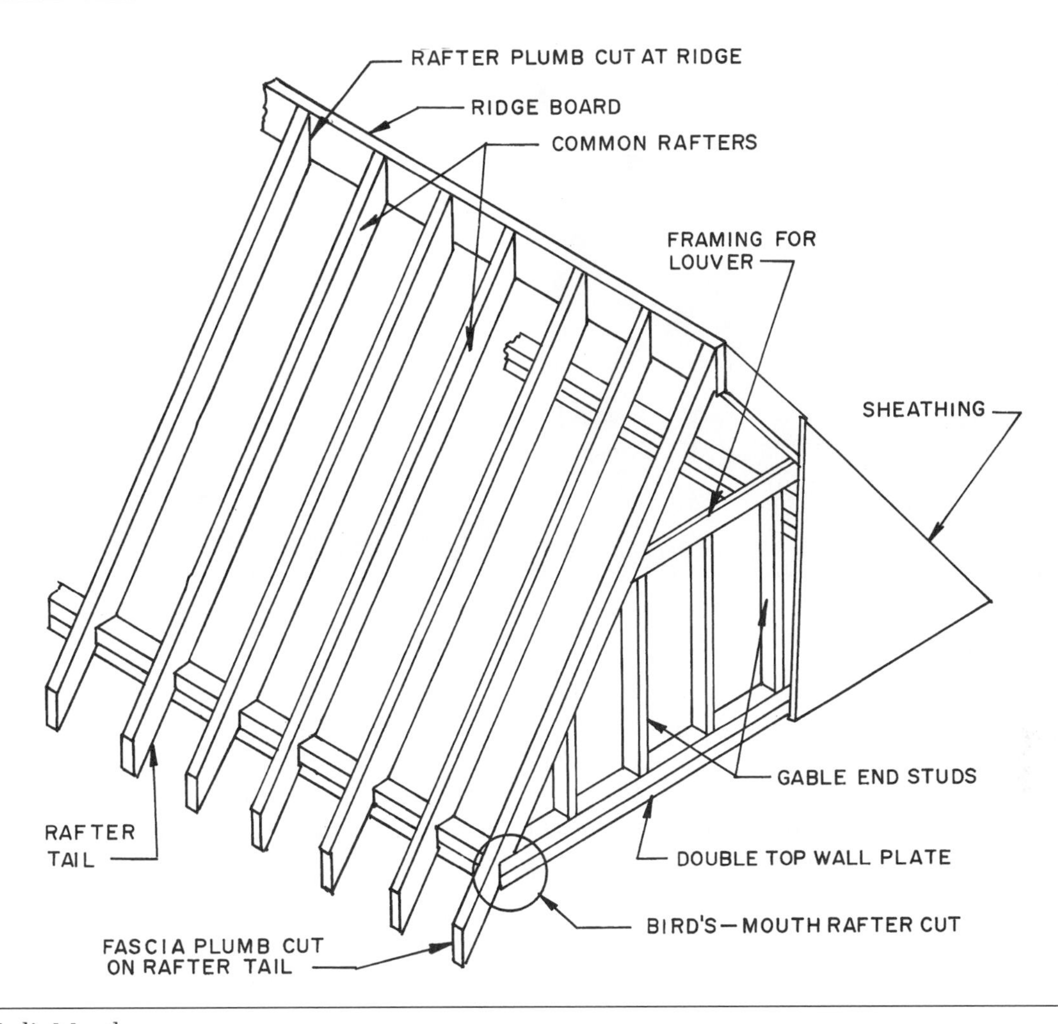

Figure B-7 Bird's Mouth.

Box Column a built-up column made of structural steel members that have either a hollow square or a hollow rectangular cross section

Box Girder a large steel beam that has a hollow rectangular cross section

Box Sill a type of sill used in frame construction in which the floor joists butt and are nailed to a header joist and rest on the sill

Brace a structural element used for positioning, supporting, strengthening, or restraining the members of a structural frame; permanent braces include lateral, diagonal, sway, and cross bracing

Braced Frame a timber or steel frame braced with diagonal members to provide lateral stability (see also *Bearing Wall System*)

Bracing temporary members used to support structural elements during construction

Brick a building unit made of fired clay

Brick Veneer a single thickness of brick wall facing placed over wood frame construction or masonry other than brick; veneer is not structural—it bears no weight except its own

Bridging diagonal or cross bracing between joists or trusses, used to stiffen, hold joists in place, and distribute loads

Brittle Materials materials that are weak in tension and shear strength that tend to break without bending

Buckling a type of column failure that occurs when the load exceeds the compressive strength of the material, and it deflects laterally; long, slender columns are subject to failure by buckling, whereas short, thick columns are subject to failure by crushing

Building any structure used or intended to support or shelter any use or occupancy

B

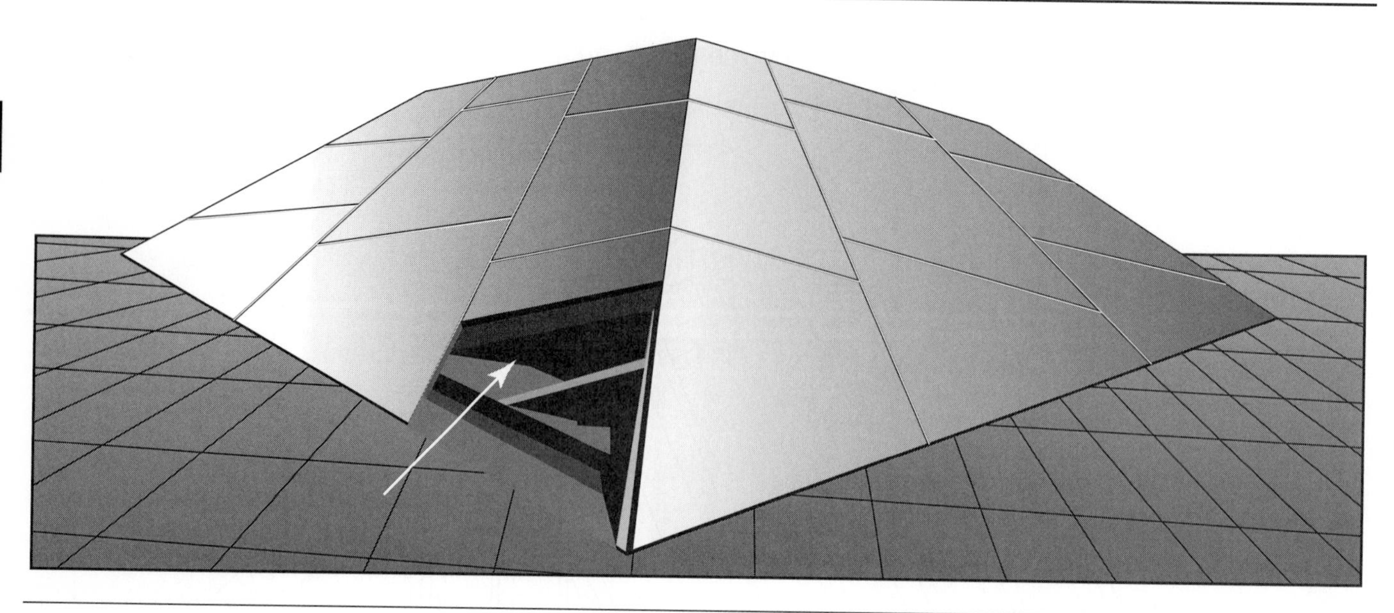

Figure B-8 Buildovers.

Building Codes standards adopted and enforced by local government agencies to regulate the design, construction, alteration, and repair of buildings to protect the public safety, health, and welfare; model codes are building codes developed by national organizations of building code officials, professional societies, and trade associations for adoption by local communities; the three major model codes are the National Building Code (NBC), Uniform Building Code (UBC), and Standard Building Code (SBC). In 2005, the International Building Code (IBC) was the first unified model code in U.S. history and was adopted by the majority of states; building codes limit the maximum height and area per floor of a building according to the type of construction and occupancy or use (automatic sprinkler systems and/or the use of fire walls may allow for greater size of the building); the intent of these provisions is to protect a building from fire and contain a fire long enough to allow people to safely evacuate

Building Envelope exterior wall coverings, as well as roof and roofing structures

Building Materials concrete, masonry, steel, nonferrous metals, stone, wood, wood panel products, plastics, glass, and glass products; their use and selection depends on functional, economic, and aesthetic considerations

Building Paper any type of various papers, felts, or similar sheet materials used in construction to prevent the passage of air or moisture

Building Types as classified by NFPA 220, there are five categories based on their structural elements' fire resistance in hours:

- **Type I (*fire resistive*)** a construction type of steel or concrete in which structural elements are protected with fire-resistive material
- **Type II (*noncombustible*)** a construction type similar to Type I, but with less fire resistance
- **Type III (*ordinary construction*)** a construction type with exterior walls of noncombustible material and interior structural materials that are typically combustible but may have steel elements; sometimes referred to as *brick-and-joist* construction
- **Type IV (*heavy timber*)** a construction type with exterior walls of masonry, brick, or stone with interior structural members made of wood not less than 6 inches by 6 inches
- **Type V (*combustible wood frame*)** a construction type whose exterior walls and interior structural members are entirely of wood

⚠ **Buildovers** concealed spaces in roof structures usually where two rooflines come together **(Figure B-8)**

Built-Up Beam a vertically laminated wooden beam made by fastening together two or more small members

Built-Up Members steel structural members that comprise sections that are riveted, bolted, or welded together

Built-Up Roofing used to cover roof decks of steel, wood, plywood, and various concrete types of varying slopes; usually applied over a vapor barrier and rigid thermal insulation; a 36-inch base sheet of glass fiber or organic roofing felt is laid; and fiberglass, asphalt, or tar-saturated felt are layered with hot asphalt or coal–tar bitumen; two-, three-, and four-ply roofing with varying felt overlaps are typical; a cap sheet of coated, mineral-surfaced felt goes down with a surfacing of bitumen (tar or asphalt), and then a wear course of a mineral/gravel surface is last to stiffen the membrane and resist wind blow off **(Figure B-9)**

Butler Building trade name for a type of building erected with prefabricated steel frame and panels

Butress a projecting structure of masonry or wood that supports or gives stability to a wall or arch; the location of the buttress will indicate the point where roof trusses or girders are supported by a bearing wall

Butt the end of a structural member

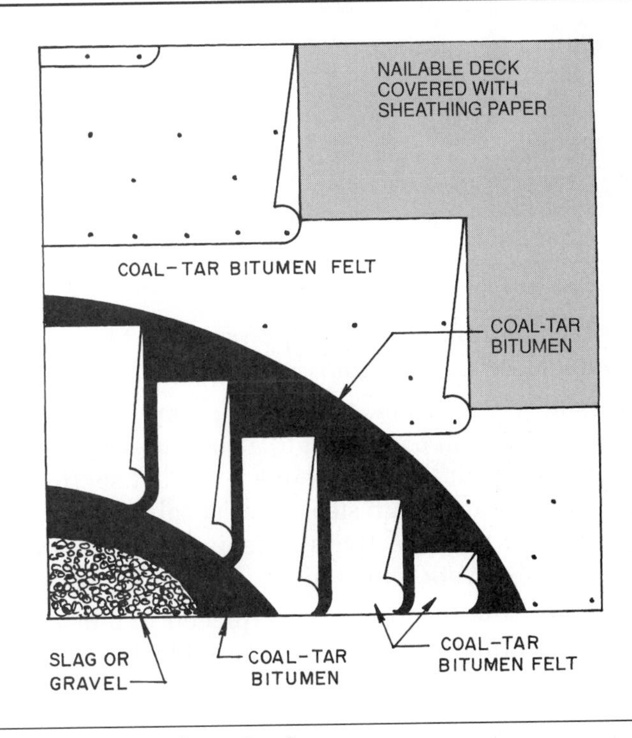

Figure B-9 Built-up Roofing.

Cable a stranded wire rope used for tensioning concrete (see also *Tendon*)

Caisson cast-in-place concrete piers formed by a shaft in the earth connected to a bearing stratum and filled with concrete; may have a steel H-section or may be reinforced with steel to resist column loading; also refers to a protective sleeve used to keep water out of an excavation for a pier

Camber a low vertical curve placed in a beam or girder to counteract deflection caused by loading

⚠ **Canopy** an overhanging roof or shelflike part of a building; suspended rods and/or beams are sometimes used, which are susceptible to failure under fire conditions **(Figure C-1)**

Cantilever a projecting beam or slab supported at one end; under fire conditions, cantilevered balconies on trusses or wooden I-beams must be monitored

Carriage sides of the stairs that support the treads, also called *stringers*

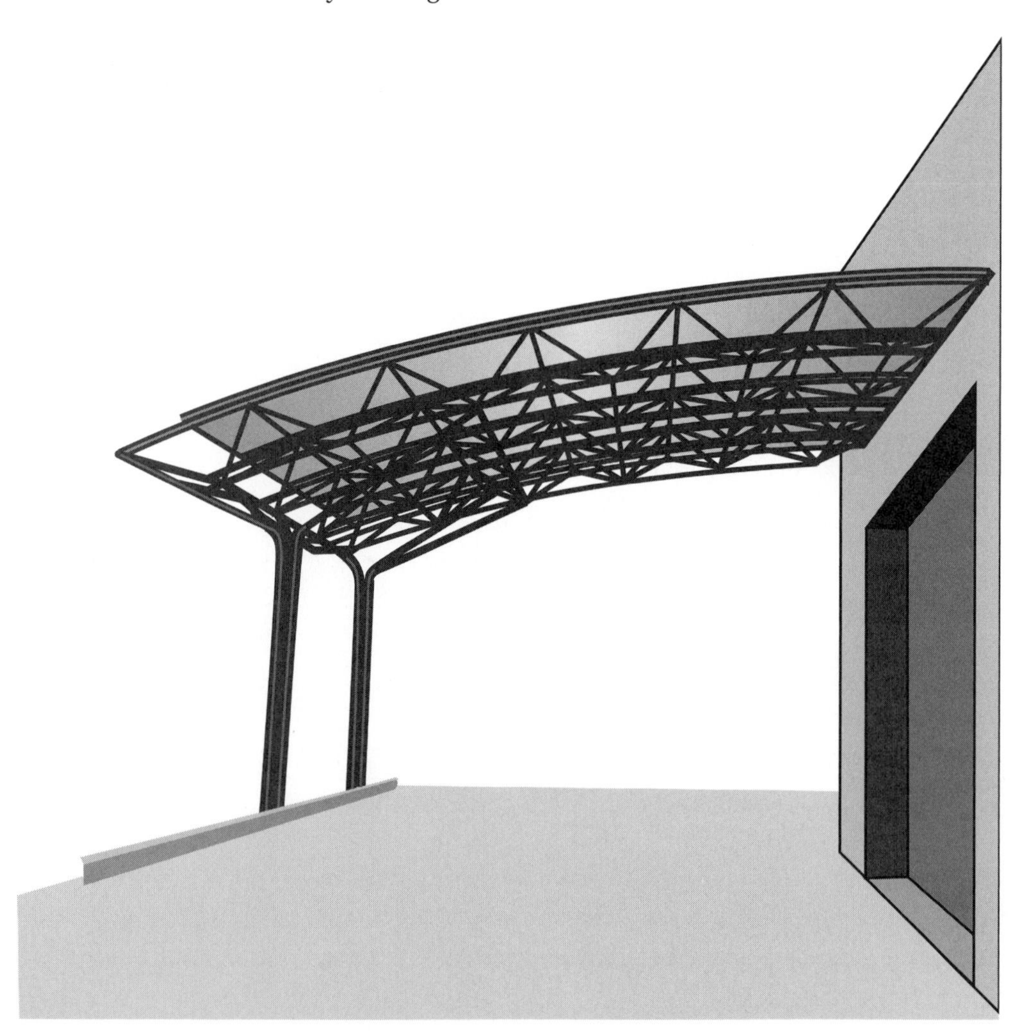

Figure C-1 Canopy.

Casing molding, usually made of wood, used around windows and doors to conceal, finish, and protect the joints between door and window frames and the surrounding wall surface

Cast-in-Place Concrete concrete formed into the desired shape where it is to remain in a structure; foundations, slabs on grade, pilings, beams, and one- and two-way slabs are common examples; cast-in-place concrete beams are almost always formed and placed along with the slab they support **(Figure C-2)**

Cast Stone a precast building stone manufactured from Portland cement and concrete, used as a trim, veneer, or facing on or in buildings or structures

Cavity Wall a hollow wall consisting of two wythes or sections of masonry units tied together with steel ties or masonry trusses

Ceiling the overhead inside lining of a room

Ceiling Joist beams that generally run the width of the building; the sizing of the ceiling joist is based on the span, spacing, load, and type of material and design **(Figure C-3)**

Cell hollow area in block

Cellular Decking a corrugated sheet welded to a flat steel sheet that forms a series of spaces or raceways for electrical and communication wiring **(Figure C-4)**

Cement a calcined mixture of clay and limestone, finely pulverized and used as an ingredient in concrete and mortar; Portland cement is commonly used, but five types of cement are available: Type I, *normal,* for general construction; Type II, *moderate,* for more resistance to sulfates, also generates less heat of hydration; Type III, *high early strength,* faster curing for earlier strength; Type IV, *low heat,* used in the construction of large concrete structures, where heat buildup can be damaging; and Type V, *sulfate resisting,* which means less need for gypsum. Air-entraining Portland cement used in Types I, II, and III is designated by the suffix A, such as Type IA

Cementitious Roof Planks noncombustible planks manufactured with Portland cement, lightweight aggregate, an aerating compound, and galvanized, welded wire-fabric reinforcement; may also consist of wood fibers that are bonded under pressure with cement; these structural planks can be used to span wood or steel roof framing and serve as roof sheathing or as permanent formwork for a concrete slab; thickness ranges from 2 to 4 inches with spans from 3 to 12 feet, and lengths vary from 5 to 12 feet with widths of 24, 30, and 48 inches

Central Heating Plant environmental heating equipment that directly utilizes fuel to generate heat in a medium for distribution, by means of ducts or pipes, to areas other than the room or space where the equipment is located

Ceramic Insulators a pre–1930s-style electrical wiring system utilizing a ceramic insulator to

C

Figure C-2 Cast-in-Place Concrete.

Figure C-3 Ceiling Joist. *(Courtesy of Pat McAuliff.)*

Figure C-4 Cellular Decking.

Figure C-5 Chase.

separate wiring; a plastic-wrapped wire system, such as Romex, is typically used today **(Figure K-2)** (see also *Knob and Tube Wiring*)

CFR Code of Federal Regulations

Champher to cut off the corners of a timber to retard ignition

Change of Use a change in the purpose or level of activity within a structure that involves a change in the application of the requirements of the building code

Channels steel structural components that have a square, U-shaped cross section

Chase a vertical or horizontal space in a building used to route pipes, wires, ducts, or other utility or mechanical systems; a channel or groove in a wall for pipes or wires to pass vertically between floors **(Figure C-5)**

Chipboard an engineered wood product that utilizes wood chips glued together to make flat sheets; sometimes used for the floors of mobile homes, and may have a tendency to weaken during fire-ground activities

Chord the main top or bottom structural member of a truss to which the diagonals are attached; either of the two principal members of a truss, extending from end to end or connected by web members **(Figure C-6)**

Cinder Block a masonry building unit made from cement with cinders used as aggregate

Cladding exterior finish or skin of a structure; prefabricated wall panels of aluminum, precast concrete, marble, stone, or enamalized steel are common, but metal cladding is primarily used to cover industrial-type buildings; it is typically 3 feet wide and spans vertically between horizontal steel girts spaced 8 to 24 feet apart

Clapboard a wood siding board that is applied horizontally; these boards are tapered in thickness, with the thick lower edge of one board designed to overlap the thinner top edge of the board below

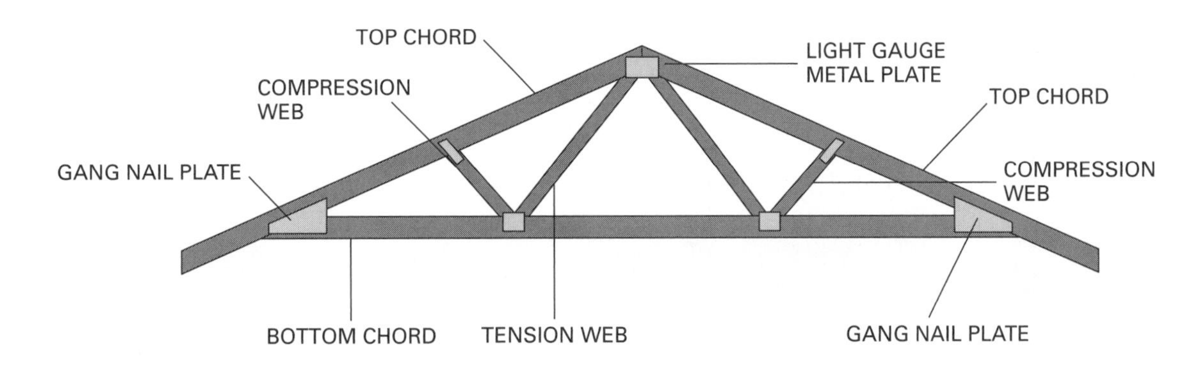

Figure C-6 Chord. *(Courtesy of* Basic Engineering for Builders, *by* Max Schwartz, *published by Craftsman Book Company.)*

Class-A Roof Assemblies a roof assembly or material with the highest fire resistance; may be tile, cement, or glass-fiber-reinforced shingles, effective against severe fire-test exposure (California Building Code, 2007)

Class-B Roof Assemblies a roof assembly effective against moderate fire-test exposure

Class-C Roof Assemblies a roof assembly effective against light fire-test exposure

Clear Span the span of a structural member between supports, also called the *allowable span* **(Figure C-7)**

Clerestory a windowed space that rises above lower stories to admit air, light, or both; may also refer to a raised section of roof with windows or a vertical wall between roofs of different heights, often used to provide light and/or ventilation

⚠ **Cockloft** the concealed space between the ceiling and the roof of a structure; distinguished from an attic by its low height; it is often viewed as part of a flat roof assembly

Code a systematically arranged body of rules; model code systems that are in place today include the International Building Codes (IBC), Standard Building Code (SBC), National Building Code (NBC), and International Fire Code (IFC). NFPA 5000, the Uniform Fire Code (UFC), and many state codes are used by several states and jurisdictions throughout the nation that are independent of any other entities; specification codes establish construction requirements by reference to particular methods and materials; performance codes set a requirement for performance that must be met; building codes regulate the construction of the building, and fire codes regulate the use of the building

Cold Drawn metal that is drawn through dies to form it without first adding heat to the metal

⚠ **Collapse** the failure of any portion of a structure; may be the result of shifting loads or

Figure C-7 Clear Span.

degradation by fire, age, deterioration, explosion, earthquake, wind, or other forces. In *Collapse of Burning Buildings,* Dunn describes the following types of collapses:

1. Curtain-fall wall collapse
2. Inward/outward collapse
3. Lean-over collapse
4. Lean-to floor collapse
5. Ninety-degree-angle wall collapse
6. Pancake floor collapse

7. Secondary collapse
8. Tent floor collapse
9. V-shape floor collapse

(Figure C-8) (see also *Failure and Structural Failure*)

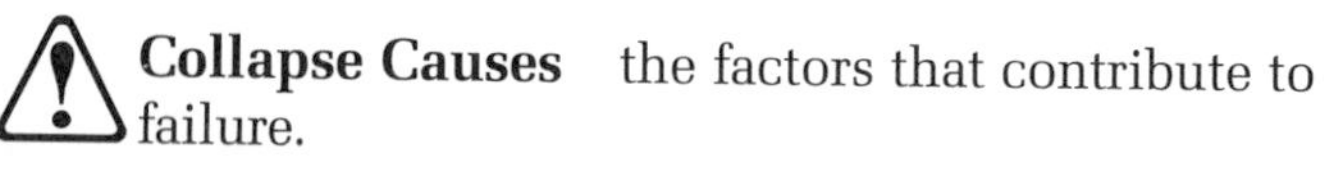

Collapse Causes the factors that contribute to failure.

1. Visible deflection, deformation, or out-of-plumb building elements

Collapse Types.

Types	Description	Fire Ground Considerations	Post Collapse Actions
Curtain-Fall Wall Collapse	Masonry that falls straight down.	Any structural failure requires defensive operations if there are no life safety/rescue issues. Establish secondary collapse zones to protect personnel. May be caused by impacts of master stream.	Announce failure of specific wall location (side A, B, C, D). Conduct PAR. Initiate RIC if appropriate.
Inward/Outward Collapse	The top collapses inward, the bottom falls outward. Timber truss roof failure may initiate secondary inward/outward collapse.	Any structural failure requires defensive operations if there are no life safety/rescue issues. Establish secondary collapse zones to protect personnel.	Announce failure of specific wall location (side A, B, C, D). Conduct PAR. Initiate RIC if appropriate.
Lean-Over Collapse	Wood-frame failure as building tilts or leans to one side.	Any structural failure requires defensive operations if there are no life safety/rescue issues. Establish secondary collapse zones to protect personnel.	Announce failure of specific wall location (side A, B, C, D). Conduct PAR. Initiate RIC if appropriate.
Lean-To Floor Collapse	One end of floor beams remains supported by a bearing wall and the other end collapses onto the floor below or collapses but remains unsupported.	Any structural failure requires defensive operations if there are no life safety/rescue issues. Establish secondary collapse zones to protect personnel.	Announce failure of specific wall location (side A, B, C, D). Conduct PAR. Initiate RIC if appropriate. Collapse creates voids that may provide survival able areas in the failure area.

Figure C-8 Collapse. *(Courtesy of* Vincent Dunn's Collapse of Burning Buildings, *Copyright Fire Engineering Books & Videos, 1988.)*

Types	Description	Fire Ground Considerations	Post Collapse Actions
Ninety-Degree Angle Wall Collapse	A monolithic masonry wall collapse at a 90-degree angle.	Any structural failure requires defensive operations if there are no life safety/rescue issues. Establish secondary collapse zones to protect personnel. Falling masonry materials may extend beyond the height of the wall.	Announce failure of specific wall location (side A, B, C, D). Conduct PAR. Initiate RIC if appropriate.
Pancake Floor Collapse	The collapse of one floor section down upon the floor below it in a flat, pancake-like configuration.	Any structural failure requires defensive operations if there are no life safety/rescue issues. Establish secondary collapse zones to protect personnel.	Announce failure of specific wall location (side A, B, C, D). Conduct PAR. Initiate RIC if appropriate.
Secondary Collapse	The collapse of one building onto another.	Any structural failure requires defensive operations if there are no life safety/rescue issues. Establish secondary collapse zones to protect personnel.	Announce failure of specific wall location (side A, B, C, D). Conduct PAR. Initiate RIC if appropriate.
Tent Floor Collapse	A floor collapse in the shape of a tent.	Any structural failure requires defensive operations if there are no life safety/rescue issues. Establish secondary collapse zones to protect personnel.	Announce failure of specific wall location (side A, B, C, D). Conduct PAR. Initiate RIC if appropriate.
V-Shape Floor Collapse	The collapse of a floor at the center of the floor beams with both ends partially supported at the outer walls.	Any structural failure requires defensive operations if there are no life safety/rescue issues. Establish secondary collapse zones to protect personnel.	Announce failure of specific wall location (side A, B, C, D). Conduct PAR. Initiate RIC if appropriate.

Figure C-8 Continued.

C

2. Failure of wood, steel, or concrete bearing elements
3. Additional live loads as a result of fire-ground actions
4. Failure of roof/floor support systems
5. Explosive impact forces
6. Prolonged fire exposure to load bearing elements
7. Secondary failure of an exposure building

(Figure C-9) (see also *Failure and Structural Failure*)

⚠ **Collapse Indicators** any movement of building parts, including sagging floors, bulging or leaning walls, columns or walls out of plumb, and/or failure of any structural member or part of the building; in wood frame buildings, an indicator would be prolonged fire impingement on structural members and/or a large volume of fire coming through windows or the roof; in commercial buildings, indicators include two or more floors on fire, unprotected steel columns and beams exposed to fire, and heavy live loads exposed to fire or water accumulation on floors **(Figure C-10)**

⚠ **Collapse Zone** the area around a building where debris will land when it falls; minimum distance is considered 1.5 times building height for low-rise buildings

Collar Beam/Tie in conventional framing, a board connecting or bracing together pairs of rafters above the plate line, which enables the structure to resist the outward thrust of the roof rafters at the eaves **(Figure R-09)** (see also *Roof*)

Collapse Causes.

Collapse Causes	Fireground Considerations
Visible deflection, deformation, or out-of-plumb building elements.	Observe and report on the condition of building (good or poor).
	Deficiencies in construction rather than the loss of fire resistance due to fire are a major reason for failure (NFPA 16th edition).
	If possible, determine a time estimate of burn time.
Failure of wood, steel, or concrete bearing elements.	Consider the load paths through unprotected steel and wood columns, beams, and bearing walls.
	Be alert to time and the exposure of fire to those elements, particularly steel beams and columns.
Additional live loads as a result of fireground actions.	Company officers to alert the IC of any retained water on floors or roofs.
	Company officers to ensure only the necessary number of firefighters are exposed as hazardous concentrated loads to floors and roofs.
Failure of roof/floor support systems.	Consider defensive operations at any structural failure.
Explosive impact forces.	Consider defensive operations at any explosion that may have impacted structural elements.
Prolonged fire exposure to load bearing elements.	Establish collapse zones in the event of secondary failure.
Secondary failure of an exposure building.	Announce collapse and have Fire Communications/Dispatch repeat announcement.
	Establish PAR.
	Initiate RIC operations if appropriate

Figure C-9 Collapse Causes.

C

Building Component	Failure Signs	Fire Conditions Predictors	Hazard Mitigators
Floors	Sagging Overloading due to increased live loads (water, firefighters)	Fire on two or more floors in unprotected or combustible frame walls	Consider defensive operations. Establish collapse zones.
Walls	Leaning Bulging Out of plumb	Fire burning combustible wood framed load-bearing walls	Consider defensive operations. Establish collapse zones.
Roofs	Sagging Overloading due to increased live loads (water, firefighters)	Fire through roof in combustible wood framed roof support systems	Consider defensive operations for large area roof area fires or ventilation at uninvolved areas. Establish collapse zones. Anticipate localized roof failure for combustible light frame wood trusses. Consider wall failure for roofs with unprotected steel trusses.
Columns	Out of plumb Bending or buckling	Fire impingement on unprotected columns	Consider defensive operations. Establish collapse zones. Anticipate localized failure due to change in load paths. Cool unprotected steel columns from safe positions.
Beams	Warping Sagging Twisting Undesigned un-levelness	Fire impingement on unprotected beams (particularly steel)	Consider defensive operations. Establish collapse zones.
Trusses	Warping Sagging Twisting Undesigned un-levelness	Fire impingement on unprotected trusses (particularly steel and light-weight wood trusses)	Consider defensive operations. Establish collapse zones. Anticipate localized roof failure for combustible light frame wood trusses. Consider wall failure for roofs with unprotected steel trusses.

Figure C-10 Collapse Indicators/Signs.

C

Column a structural member that transmits a compressive force along a straight path in the direction of the member, typically a vertical supporting member primarily designed to support axial, compressive forces of roofs and floors; columns transfer the building weight to the foundation, so the loss of a load-bearing column will have severe consequences; the most efficient shape for a column is one that distributes the material equally around the axis as far as possible from the center. Stability factors and variables include the material, height, connections, shape, and whether it is braced or unbraced, hollow or solid; the load capacity of a column is also determined by its slenderness ratio: engineers identify concrete columns as long, short, or pedestal based on the ratio of length to; concrete columns have vertical and horizontal reinforcing steel, called *compression bars* for vertical reinforcement; failure can result from crushing or buckling by excessive stresses as well as by how it is loaded and the types of connections used

Column Cap the metal plate on the top of a support column

Combustible of, or pertaining to, a material capable of igniting and burning

Common Bond an arrangement of brickwork in which the majority of the courses are stretchers interspersed with headers at certain intervals

Common Rafter a rafter that extends from a wall top plate to either a ridge board or a ridge beam; supports the sheathing and covering of a roof

Compartmentation the dividing of a larger space into smaller spaces by smoke-stop partitions to limit the combustible space; common attics and/or concealed spaces are compartmentalized with framing and gypsum wallboard, but compartmentation cannot be relied upon to confine fire: alterations/modifications by tenants, maintenance personnel, and construction workers may limit the effectiveness of compartmentation due to holes cut in walls

Component a part of a whole

Composite built up of different parts, pieces, or materials

Composite Column a structural steel column thoroughly encased in concrete and reinforced with both vertical and spiral reinforcements

Composite Construction construction in which one or more materials are used in the structural elements; examples include the use of concrete exterior columns and steel interior columns, bricks and blocks tied together, and steel columns with wood glulams **(Figure C-11)**

Figure C-11 Composite Construction. *(Courtesy of Kathleen Siegel.)*

Composite Decking a metal deck that serves as tensile reinforcement for the concrete slab to which it is bonded with embossed rib patterns; typical corrugation depths range from 1.5 to 3 inches and spanning 4 to 15 feet

Composite Material two materials combined to take advantage of the best characteristics of each, such as reinforced concrete (steel and concrete); failure or damage to one of the elements will be the weak link in the composite material's overall strength

Composite Structural Element two different materials combined in a structural element; reinforced concrete utilizing steel and concrete is one example

Composite Wall a wall made up of two or more different materials, such as brick and concrete block, designed to react as one under load

Composition Shingle a roof covering used on some sloped roofs, made from asphalt and structural fibers bonded together and covered with a protective granular coating; most composition shingles have tabs with self-sealing adhesive or locking tabs that make them wind-resistant: inorganic fiberglass-based shingles have excellent fire resistance (UL Class A), and organic felt-based shingles have moderate fire resistance (UL Class C); also known as *prepared roofing* (see also *Fiberglass Shingle*)

Compound Columns structural steel columns encased in concrete at least 2.5 inches thick and reinforced with wire mesh; steel columns may also support wood beams

Compression a pushing force that puts inward pressure on an object or material

Compression Load a weight or force tending to compress a supporting member; to push materials together

Compressive Strength the ability of a material to resist a heavy compressive load

Concave a surface that is curved or dished inward

⚠ **Concealed Space** void areas that are covered and yet allow for fire spread to occur undetected; such spaces are responsible for the loss of many large structures because hidden voids were not opened up early enough; NFPA 5000 requires that any concealed space in which building materials having a flame-spread index greater than Class A will be draft-stopped within specific requirements; in addition, concealed spaces constructed of combustible materials shall be fire blocked with specific requirements **(Figure C-12)** (see also *Void*)

Figure C-12 Concealed Space. *(Courtesy of Kathleen Siegel.)*

Concentrated Load a load that is applied at one point or over a small area, such as an air conditioning unit on a roof

Concrete a mixture of cement, sand, and aggregate that solidifies through chemical reaction into a structural material with good compressive strength when mixed with the proper proportion of water; although strong in compression, steel reinforcement is required to handle tensile and shear; concrete structures are relatively low in cost and are inherently fire-resistant; concrete is heavy (150 pounds per cubic foot), it requires forms for to be poured in place, and it has a lack of ductility that makes concrete structures vulnerable to damage from earthquakes; admixtures may be used to change its load-carrying capabilities and strength; insulating concrete has low thermal conductivity and is lightweight (60 pounds per cubic foot)

Concrete Beams reinforced concrete beams are designed to act together with longitudinal and web reinforcement in resisting applied forces

C

Concrete Block a masonry unit, usually of hollow construction, that has been cast from concrete; these building bocks come in a wide variety of shapes, are made with Portland cement and aggregates, and are also known as *concrete masonry units, or CMUs* (see also *Masonry Block*)

Concrete Bond the cementing together of two or more concrete surfaces

Concrete Columns designed to act together with vertical and lateral reinforcement in resisting applied forces; reinforced concrete columns are usually cast with concrete beams and slabs to form a monolithic structure but can also support a grid of timber or steel beams; steel connectors are required to support and anchor wood or steel beams to concrete columns; corbels may be cast into precast concrete columns to provide additional bearing for floor and roof slabs; the most frequently used section for columns is the wide-flange (W) shape; it is suitable for connections to beams in two directions, and all of its surfaces are accessible for making bolted or welded connections; other steel shapes used for columns are round pipes and square or rectangular tubing

Concrete Construction a structure utilizing concrete as its principal material; typical concrete constructed buildings include precast, poured-in-place, and prestressed treatments; precast concrete elements are cast and allowed to harden and then placed in position for their desired use (beams, roof planks, etc); poured-in-place concrete is poured in the location where it is going to exist utilizing formwork; prestressing creates greater strength in concrete by utilizing tendons that are tensioned (includes pretensioning and posttensioning)

Concrete Floor System a horizontal plane of concrete that supports live and dead loads and transfers those loads to beams, columns, or load-bearing walls; typical concrete floor systems are either slabs of reinforced concrete to span one or both directions (one- or two-way slabs) of a structural bay or precast concrete planks. Reinforced concrete beams are designed to act together with longitudinal and web reinforcement in resisting applied forces; cast-in-place concrete beams are almost always formed and placed along with the slab they support; the depth of a floor system is directly related to the size and proportion of the structural bays it must span and the strength of the materials used; lightweight concrete floors are often used in multiple story buildings and must not be mistaken as a slab resting on the ground **(Figure C-13)**

Figure C-13 Concrete Floor System.

Concrete Formwork used for columns, floors, and walls for cast-in-place concrete construction; fresh concrete must be shaped and supported by formwork until it cures and can support itself, the framework and bracing must be able to maintain the position and shape of the forms until the concrete sets; the contact surfaces of forms are typically coated with a parting compound (oil, wax, or plastic) to aid in their removal **(Figure C-14)**

Figure C-14 Concrete Formwork.

Concrete Masonry a modular material known as *block,* used for foundations and walls; small (8-inch) units allow for easy handling and do not require formwork; mortar joints, steel vertical reinforcement, and grout filling are used; masonry bearing walls may support concrete, steel, and wood floor and roof systems; embedded anchor bolts, steel-bearing plates, metal wire ties, twisted steel straps, reinforced bond beams, and/or steel dowels secure those floor and roof systems to the walls and foundation

Concrete Slab flat, horizontal concrete pour

Concrete Structures structures constructed with concrete that include cast-in-place concrete, precast concrete, or a combination of the two; most concrete structures have steel reinforcement: lateral-force–resisting system types include moment-resistant frames, sheer walls (cast-in-place), and in-fill masonry (concrete frame structures); roofs and floors serve as horizontal diaphragms, distributing lateral loads among vertical resisting elements; floor framing systems include beams and one-way slabs, beams and two-way slabs, beams and one-way joists and slabs, waffle slabs, and flat slabs

Concrete Wall a concrete wall that has been poured in place into forms and allowed to set up in its permanent location

Concrete Wall Systems noncombustible construction that may be in the form of a structural frame, such as a rigid frame or as bearing walls. Structural frames can support and accept a variety of nonbearing or curtain wall systems; concrete and masonry bearing walls rely on their mass for their load-carrying capability; although strong in compression, concrete and masonry require reinforcement to handle tensile stresses.

Conduit a channel or pipe used for conveying water; a tube or duct for enclosing electric wires or cable

Connection an element of a structure that joins members together so that loads and forces may be transferred from one structural element to the next; structural members must be adequately interconnected for the proper force transfers to be made; the performance of the structural system as a whole depends upon the types of joints and connections used; structural elements can be joined to each other in three ways: *butt joints, overlapping joints,* or *molded/shaped joints* form a structural connection; these joints may take the form of *a point* (pinned, like nails, bolts, or rivets); *a line* (linear like a weld) or *a surface* (glued, surface to surface); rigid or fixed joints maintain the angular relationship between the joined elements, restrain rotation, and provide both force and moment resistance, like welded connections and reinforced concrete; structural failure is often associated with deficiencies in construction, often at joints and connections

Connector a device to join two objects together; pinned and rigid-framed are the two major groupings; pinned connectors include nails, rivets, and bolts; rigid-framed connectors typically join steel or concrete elements; for rigid-frame concrete connectors, grade beams and monolithic reinforced concrete elements are used; light-frame wood construction and heavy timber construction are highly dependent upon the quality of connectors and connections; structural members, shear walls, and floor and roof diaphragms must be adequately interconnected for the proper force transfers to be made. Connections and connector failures often prove to be the weak point in wood and timber frame construction due to shear failure. Timber connections are an integral component of structural integrity and have contributed to failures in timber structures **(Figure C-15)**

Construction Adhesive a strong adhesive used to hold structural and/or nonstructural members together

Construction Types building codes classify the construction of a building according to the fire resistance of its major components, which include structural frame, exterior bearing and nonbearing walls, interior bearing walls, permanent partitions, floors, ceiling and roof assemblies, and stairways and shaft enclosures; protected construction is required to be of 1-hour fire-resistive construction throughout; unprotected construction has no requirements for fire resistance except for shaft and exit enclosures, or where the building code requires protection of exterior walls due to their proximity to a property line; Type I buildings are of noncombustible construction and have a structural framework of steel, reinforced concrete, or reinforced masonry; Type II buildings are similar to Type I buildings except for a reduction in the required fire-resistance ratings of building elements; Type III buildings have noncombustible exterior walls and a structural framework of steel, concrete, masonry, or wood; Type IV buildings have noncombustible exterior walls, an interior structural framework of solid or glue-laminated timber of specified minimum sizes, and floors with no concealed spaces; Type V buildings have elements whose fire-resistance ratings are specified in the building code **(Figure C-16)**

Contents furnishings, decorations, stock, or goods placed inside a structure

Continuous without a break or termination

Continuous Beam a beam supported at three or more points

Continuous Casting a process of pouring concrete without allowing it to harden

Continuous Load Path defined by NFPA 5000 as transferring all gravity, seismic, and wind loads from the roof, wall, and floor systems to the foundation

Continuous Ridge Vent a vent into an attic located along the ridge of a roof

Control Joint a joint between sections of a structure designed to permit differential movement between the structures; such joints relieve stresses in the sections, which minimizes damage and may be found in concrete slabs and gypsum-board wall installations

Conventional Framing solid two-by dimensional wood lumber for roof and floor framing as compared to engineered, lightweight trusses and I-beams

Convex a surface that curves outward

Copestone the finishing piece of stonework along the top of a structure

Coping the cap on a masonry wall or other structure that prevents water from penetrating through the top; may also refer to a low decorative wall above the roof line

Corbel a projection from the face of a wall or column that usually projects more with each succeeding course of masonry units; may be decorative, as on top of a parapet wall, or may be used on the inside of a masonry wall to support a beam

Cored Section a cylindrical hole in precast concrete

Corner the junction of two planes or surfaces

Corner Bead a formed piece of metal designed to protect the outside edge of a wall corner from wear

Cornice a decorative horizontal projection that crowns a wall or finishes the eaves of a building; may also refer to the junction formed by the eaves and walls of a building **(Figure C-17)**

Corridor a narrow hallway or passageway with rooms or apartments opening onto it; an enclosed exit access component that defines and provides a path of egress travel to an exit

Corrosion the quality of a substance that can cause the chemical deterioration of a material

Corrugated a surface that is composed of a series of alternating, rounded ridges and valleys; for roofing

C

Connector Types		Failure Issues	Hazard Mitigators
Pinned	Nails	Shear forces Poor workmanship	Identify and report connector signs of weakening, deformation or exposure by fire and location(s).
	Rivets	Shear forces Poor workmanship	Identify and report connector signs of weakening, deformation or exposure by fire and location(s).
	Bolts	Shear forces Poor workmanship	Identify and report connector signs of weakening, deformation or exposure by fire and location(s).
	Screws	Shear forces Poor workmanship	Identify and report connector signs of weakening, deformation or exposure by fire and location(s).
Rigid Frame			
	Glue-laminated wood timbers	Unprotected Long burn time	Identify and report connector signs of weakening, deformation or exposure by fire and location(s).
	Steel	Unprotected Unknown weld quality	Identify and report connector signs of weakening, deformation or exposure by fire and location(s).
	Reinforced concrete	Exposed steel reinforcement due to spalling of concrete	Identify and report connector signs of weakening, deformation or exposure by fire and location(s).

Figure C-15 Connector Types.

it may be aluminum, galvanized steel, and fiberglass or reinforced plastic and structural glass

Course a horizontal layer of individual masonry units

Court an open, uncovered, unoccupied space unobstructed to the sky

Cove Molding a molding with a concave face designed to fit into an interior corner

Crawl Space the space within the foundation perimeter under a building's flooring that allows access to plumbing pipes and other systems

Cricket a small peaked structure on a roof for diverting rainwater around a projection, such as a chimney on a sloping roof (see also *Saddle*)

C

	TYPE I		TYPE II		TYPE III		TYPE IV	TYPE V	
BUILDING ELEMENT	A	B	A[a]	B	A[b]	B	HT	A[a]	B
Structural frame[a]	3[c]	2[c]	1	0	1	0	HT	1	0
Bearing walls									
Exterior[d]	3	2	1	0	2	2	2	1	0
Interior	3[c]	2[c]	1	0	1	0	1/HT	1	0
Nonbearing walls and partitions exterior	See Table 602								
Nonbearing walls and partitions									
Interior[e]	0	0	0	0	0	0	See Section 602.4.6	0	0
Floor construction Including supporting beams and joists	2	2	1	0	1	0	HT	1	0
Roof construction Including supporting beams and joists	1½[b]	1[b,f]	1[b,f]	0[b,f]	1[b,f]	0[g,f]	HT	1[b,f]	0

For SI: 1 foot = 304.8 mm.

[a] The structural frame shall be considered to be the columns and the girders, beams, trusses, and spandrels having direct connections to the columns and bracing members designed to carry gravity loads. The members of floor or roof panels that have no connection to the columns shall be considered secondary members and not a part of the structural frame.

[b] Except in Groups F-I, H, M and S-1 occupancies, fire protection of structural members shall not be required, including protection of roof framing and decking where every part of the roof construction is 20 feet or more above any floor immediately below. Fire-retardant-treated wood members shall be allowed to be used for such unprotected members.

[c] Roof supports: Fire-resistance ratings of structural frame and bearing walls are permitted to be reduced by 1 hour where supporting only a roof.

[d] Not less than the fire-resistance rating based on fire separation distance (see Table 602).

[e] Not less than the fire-resistance rating required by other sections of this code.

[f] In all occupancies, heavy timber shall be allowed where a 1-hour or less fire-resistance rating is required.

[g] An approved automatic sprinkler system in accordance with Section 903.3.1.1 shall be allowed to be substituted for 1-hour fire-resistance-rated construction, provided such system is not otherwise required by other provisions of the code or used for an allowable area increase in accordance with Section 506.3 or an allowable height increase in accordance with Section 504.2. The 1-hour substitution for the fire resistance of exterior walls shall not be permitted.

Figure C-16 Construction Types. *(Source: 2006 International Building Code. Copyright 2006. Falls Church, VA: International Code Council, Inc. Reproduced with permission. All rights reserved.)*

Figure C-17 Cornice.

Cripple any framing member that is shorter than usual, such as a stud above a door opening or below a windowsill

Cripple Wall a stud wall less than 8 feet in height, between the foundation and lowest framed floors, with studs not less than 14 inches long; also known as a *knee wall* **(Figure C-18)**

Figure C-18 Cripple Wall.

Cross Beam a structural beam spanning across a space

Cross Bracing one method of ensuring lateral stability in timber or steel frame construction by using diagonal members as bracing; may be rigid or cable bracing

Cross Bridging small wooden pieces placed at angles, so that they extend from the bottom of one floor joist to the top of the adjacent joist to add stability to the structural members

Crosswall a wall at a right angle to another wall

Crown Molding any ornamental molding that terminates the top of a structure or decorative feature, such as the intersection of a wall and ceiling

Curtain Board vertical board extending down from a ceiling intended to limit the horizontal spread of fire and heat **(Figure C-19)** (see also *Draft Curtain/Draft Stop)*

Figure C-19 Curtain Board.

Curtain Wall an exterior wall, typically found in high-rises, supported wholly by the steel or concrete structural frame of a building and carrying no loads other than its own weight and wind loads; panelized and prefabricated, these assemblies use diverse materials, such as glass and granite; curtain walls are subject to moisture penetration, which may

Figure C-20 Curtain Wall.

result in weakness and connector failure under fire conditions; sealants and noncorrosive anchors must be carefully selected for curtain walls and facades; a curtain wall may consist of metal frame that holds panels of concrete, stone, masonry, metal, or glass suspended in front of the structural framing. Various metal devices are used to secure a curtain wall to the structural frame of a building: these include tees, brackets, angles, plates, struts, anchors, and inserts **(Figure C-20)** (see also *Panel Wall*)

Dead Load the static load acting vertically downward on a structure and any equipment permanently attached or built in; dead loads consist of the weight of all materials of construction incorporated into the building, including but not limited to walls, floors, roofs, ceilings, stairways, built-in partitions, finishes, cladding, and other fixed service equipment

Deck/Decking a horizontal surface covering, typically metal or wood, supported by floor or roof beams. Metal decking is corrugated to increase its stiffness and spanning capability, and corrugation depth determines span; there are three major types of metal decking, topped with 2- to 3-inch concrete floor slabs: *form, composite,* and *cellular;* wood decking includes solid and laminated tongue-and-groove lumber that can span from 6 to 20 feet, depending upon its thickness; form decking serves as permanent formwork for a reinforced concrete slab, with typical depths of 9/16 to 2 inches spanning 1.5- to 12-foot lengths

Decorative Materials all materials utilized or applied over the buildings interior finish for decorative, acoustic, or other effects; does not include floor coverings or window shades

Deflection the movement of a structural element under a load, increasing with load and span, typically viewed as the downward bending of a beam due to its own weight or to an imposed load in both compression and tension; code specifies allowed values for structural members **(Figure D-1)**

⚠ **Deformation** an alteration in shape or form caused by stress, as when a load is applied, and the shape bends; in compression, the deformation takes the form of shortening, whereas in tension the deformation takes the form of elongation

Degradation deterioration or lessening in strength

Delaminate to come apart in sheets or layers; a separation of layers that were, or were intended to be, fastened together due to failure of an adhesive bond; plywood exposed to the elements will delaminate over time

Figure D-1 Deflection.

Demolition the destruction of a structure for removal

Density the mass of a substance per unit of volume of that substance

Design Load buildings are designed with a given load in mind, measured in pounds per square foot; BOCA Code specifies the minimum live-floor load design for specific types of buildings

Designer a person who plans and provides details for fabrication or construction

Detached Building a separate, single-story building, without a basement or crawl space, used for the storage or use of hazardous materials and located an approved distance from all structures (California Building Code, 2007)

Diagonal Bracing wood boards, typically 1-by-4 stock, that are let into faces of studding or cut to fit between studs to provide rigidity and structural strength

Diagonal Tension tension that results from the principal tensile stresses acting at an angle to the longitudinal axis of a beam

Diaphram a structural sheathing material applied over framing that acts as a load-bearing unit, designed to resist lateral loads. May also be referred to as a *partitioning surface,* or *stress skin,* for external applications; horizontal diaphragms may include roof sheathing or subfloors; vertical diaphragms create shear walls (vertical resisting elements) when tied together in particular ways to strengthen framing members. Also known as a *membrane*

Dimension a measurement value

Dimension Lumber lumber precut to a particular size for the building industry, usually in sizes between boards and timber, around 2 to 5 inches thick with a width of 2 inches or more **(Figure D-2)**

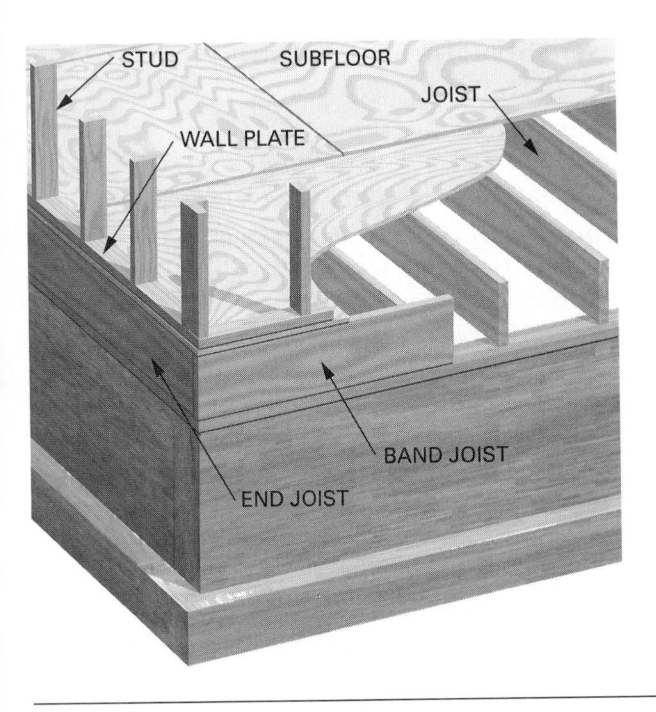

Figure D-2 Dimension Lumber.

Distributed Load load applied relatively equally over an area

Dog Iron an iron strap holding two members together and designed to fail before the members do

Dome typically a roof of various shapes; also described as a *shell*

Dormer a projecting structure built out from a sloping roof and housing a vertical window or ventilating louver

Draft Curtain/Draft Stop a material, device, or construction installed to restrict the movement of air within open spaces of concealed areas of building components, such as crawl spaces, floor/ceiling assemblies, roof/ceiling assemblies, and attics (California Building Code, 2007); it may be a partial wall of noncombustible material extending down from the ceiling as a barrier to heat flow, and it is used to compartmentalize space in large open areas as a way to minimize the mushrooming effect of heat and smoke; half-inch gypsum board, or 3/8-inch plywood is usually applied parallel to the main frame members to divide an enclosed attic space or concealed space; uniform Building Code 1-223 states that attics, mansards, overheads, and other concealed roof spaces of combustible construction need to be draft stopped; some firefighter concerns include poorly fitted sheets and penetrations by plumbing, wiring, and ductwork **(Figure C-21)** (see also *Curtain Board*)

Drip Edge metal edging placed around the edge of a roof prior to installing the roofing material; drip edges are provided on eaves and gables of shingle roofs

Drop Panel used in concrete construction to resist shear forces near columns

Dropped Ceiling another name for a suspended ceiling

Dry Rot deterioration of wood by fungi, usually as a result of exposure to a constant source of moisture

Drywall a system of interior wall finish using sheets of gypsum board and taped joints; panels are normally 4 feet wide and 8, 10, 12, 14, or 16 feet in length; there are six standard thickness available, ranging from ¼ to 1 inch, to meet the requirements of different conditions; gypsum, the formal name for drywall, is naturally fire resistant; type X panels increase the fire resistance by adding glass fibers to the core, and exterior sheathing panels can be used for indirect exposure to the weather on vertical surfaces

Duct thin-wall sheet metal or lightweight tubular conduit used for conveying air at low pressure

Ductile/Ductility a material property that allows the material to be deformed or worked into different shapes

Ductile Materials materials that tend to deform before breaking

Ductwork an installed duct system

Dwelling a structure in which people live; typically a building that contains not more than two dwelling units

Dynamic moving, changing

Eave the edge of a roof that projects beyond the exterior walls of a structure

Eave Course the first course of shingles along the eave line

Eave Flashing roofing material that is laid along the eave line and extended up the roof under the shingles

Eave Line the overhanging edge of the eave of a roof

Eccentric rotating off-center or out of round; may be oval or elliptical

Eccentric Loads a load perpendicular to the cross section of the supporting element that does not pass through the center of the mass

Edge the narrow surface of an object perpendicular to its face

Edge Joist the joist at the inside edge of a building wall (see also *Rim Joist*)

Edge Venting regularly spaced vents around the eave line of a roof that provide ventilation to the attic space under the roof

Edifice a large structure or building

Egress a safe and adequate access from any point in a building to protected exits leading to a place of refuge; the three components of an egress system consist of *exit access, exits,* and *exit discharge*

Elastic the ability of concrete, metal, or wood members to return to their original size and shape after deformation under load; *elasticity* refers to the relative stiffness of a material

Elastomeric the ability of a material to return to its original shape and after being distorted

Element an architectural or mechanical component of a building, facility, space, or site (California Building Code, 2007)

Elevation in measurement, the height as a point above sea level; in drafting, the drawing or view of any of the vertical sides of a structure

Elevator a platform or enclosure for moving people and/or commodities from one level of a structure to another

Elongate to make longer when under load or expansion due to temperature increases

Enclose to close in, or surround; enclosures create voids and concealed spaces that may allow fire to go undetected

Enclosed Structure a building with limited openings for entrance or egress

Enclosure System the shell or envelope of a building, consisting of the roof, exterior walls, windows and doors (see also *Envelope*)

Enclosure Wall a wall that separates a vertical opening for a stairwell, elevator, or shaftlike opening that connects two or more floors; may also be an exterior wall that is not a bearing wall

End-Bearing Pile Or Piling a piling that is loaded on the end so that it acts as a column

End-Grain the grain pattern exposed on the end edge when a board is cut across the grain

End-Matched lumber with tongues and grooves at the ends

Endothermic a reaction that absorbs heat

End-Wall Column a vertical structural member in the end wall of a building

Engineering the application of science and mathematics to understand how the properties of matter and the sources of energy in nature can be used to benefit man

Engineered Wood Structural Members engineered wood composite members fabricated off-site, created by gluing and pressing together veneers; typical examples include sheets, beams and columns including wood I-beam (laminated flanges with a particle web), laminated veneer lumber (LVL) and beams (Paralam trade name), plywood, chipboard, wood trusses, flitch plate girders, and wood and plastic composite roof panels **(Figure E-1)**

English Bond a brickwork pattern using alternated courses of headers and stretchers

Entasis a slight and gradual curved tapering of a column toward its center to add visual appeal

Figure E-1 Engineered Wood Structural Members.

Entrance the passageway into a building, also called an *entryway*

Envelope the outside, weather-protective coverings of the roofs, curtain walls, facades, and other parts of structures; an exterior wall envelope protects a building's structural members from the detrimental effects of the exterior environment; modern construction relies on many sealants for the integrity of the building envelope (see also *Enclosure System*)

Expoxy Resin an adhesive used to bond wood, metal, glass, and masonry

Equilibrium a state of balance in which opposing influences or forces are equalized or remain in check

Erect to construct or raise

Erection the act or process of constructing a building or structure

Essential Facilities buildings and other structures that are intended to remain operational in the event of extreme environmental loading from wind, snow, or earthquakes; many fire stations are considered essential facilities

Existing Structure a structure erected prior to the date of adoption of the appropriate code or one for which a legal building permit has been issued

Exit a passageway out of a structure, typically a continuous, unobstructed means to egress to a public throughway; building codes specify the maximum distance of travel to an exit according to a building's use, occupancy, and degree of fire hazard; all exit stairs and enclosures require a 2-hour fire-resistance rating

Exit Discharge that portion of a means of egress between the termination of an exit and a public throughway

Expansion Bolt a type of masonry-anchoring device consisting of a bolt and expandable sleeve

Expansion Joint strips of material used separate units of concrete to give the concrete room to expand with temperature changes, also used between gypsum wallboard panels

E

Exposed Aggregate a horizontal concrete surface in which decorative aggregate has been embedded after the concrete is poured but before it has completely set

Exoskeleton exterior load-bearing walls **(Figure E-2)**

Figure E-2 Internal Skeleton Frame with Non-bearing Exterior Curtain Walls.

Exposure open to the elements; a part not hidden from view

Extend to lengthen or stretch

Extension a projection

Exterior Door a door designed to be exposed to the elements on at least one side

Exterior Sheathing gypsum drywall panels and engineered wood panel products designed to be installed under finished siding materials on the vertical surfaces of exterior building walls

Exterior Soffits an enclosed overhang creating a hidden space **(Figure S-04)** (see also *Soffit*)

Exterior Surface a weather-exposed surface

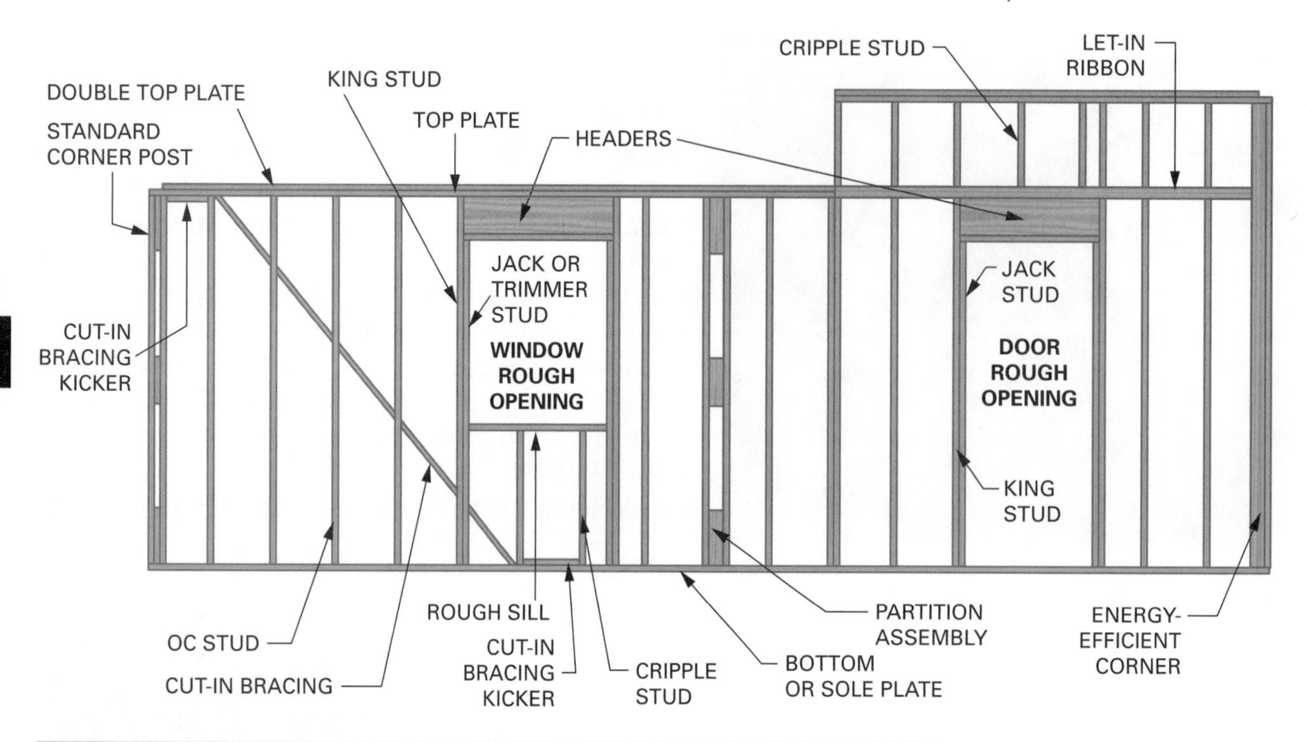

Figure E-3 Exterior Wall.

Exterior Wall a wall in which one side forms the outside exposed surface of a building **(Figure E-3)**

Exterior Wall Covering a material or assembly of materials applied on the exterior side for the purpose of providing a weather-resisting barrier, insulation, or for aesthetics, including veneers, siding, exterior insulation and finish systems, and trims (2007 California Building Code)

Fabricate to make from raw materials or assemble from parts, such as the roof trusses used in roof framing; trusses are made from structural members cut to size and joined together using metal plates with integral fasteners and are usually assembled off-site and brought to the building site for framing installation

Facade the front of a building or the nonstructural covering on the face of a building, also known as a *false front;* facades are never load bearing and are considered nonstructural; precast concrete facade panels on multistory buildings have failed after earthquakes; masonry facades, including brick or stone, are placed on structural frames for weather protection of the enclosed space, for appearance and as a protection for the structural elements against fire and climate changes **(Figure F-1)** (see also *False Fronts*)

Face an exposed surface

Face Brick brick that forms the outside surface of a wall

Faced Masonry a masonry structure with backing and facing masonry of different materials, such as brick on a concrete wall

Faced Wall a structural wall with a veneer covering

Face Veneer the outer layer of veneer on plywood

Facing a finished wall surface; smoothing and finishing a surface

Factor of Safety the ratio of the expected strength of the structure to the expected imposed forces on the structure

Factory Edge the long edge of a wallboard panel that comes from the factory covered with paper

False Mansard a mansard-type roof, typically attached to a structure like a strip mall, that creates a concealed void space that acts as a horizontal chimney for heat, smoke, and fire to travel through; these are often held in place with lightweight fasteners (nails, gusset plates) that have failed and injured and killed firefighters **(Figure F-2)**

Falsework temporary structural supports used during construction and then removed

Figure F-1 Façade.

Failure the condition of becoming incapable of performing a particular function; "An unacceptable difference between expected and observed performance" (American Society of Civil Engineers); failure may include structural collapse, distress, excessive deformation, and premature deterioration of materials, leaking roofs, and facades; partial or total failure occurs when the structure cannot redistribute the load it carries; in steel framing

Figure F-2 False Mansard. *(Courtesy of Kathleen Siegel.)*

under nonfire conditions, most failures occur due to stresses in the bolted or welded connections; when a steel beam is stressed beyond its design capability, or heated to its yield point, it will deflect, twist, or buckle **(Figure F-3)** (see also *Collapse, Structural Failure*)

Fascia the broad, flat surface of the outer edge of a roof fastened to the outer ends of roof rafters

Fastener any mechanical means of holding two materials together, such as a nail, screw, or slip

Felt a sheet material made using a fiber mat that has been saturated and topped with asphalt

Fiberboard sheet material used in many construction applications, such as floor underlayment, wall sheathing under finish walls, and cabinet walls that

CAUSES OF STRUCTURAL FAILURE

FAILURE FACTOR	CAUSE
Fire Destruction to Structural Elements	♦ Lengthy fire impingement
Poor Condition of Building	♦ Age ♦ Abandonment ♦ Exposure to weather
Overloading/Exceeding the Capacity of Structural Elements	♦ Water Accumulation ♦ Excessive tensile forces in web members of trusses, columns and arches ♦ Excessive shear forces between/within elements/materials
Improper Alignment and Lack of Structural Stability of Structural Elements and/or Structural Systems	♦ Walls, columns not plumb ♦ Alterations by unqualified workers ♦ Floor and roof support system load path shifts
Connector Failure	♦ Roof to wall ♦ Floor to wall
Construction Features	♦ Light weight wood construction materials and engineering ♦ Unreinforced masonry walls ♦ Undivided and/or unprotected truss voids
Explosion	♦ BLEVE ♦ Backdraft ♦ LPG/natural gas ignition ♦ Terrorist action
Impact By Master Stream	♦ Lateral impact loads

Figure F-3 Causes of Structural Failure.

are to be covered with a veneer made by compressing wood or other plant fibers with a binder

Fiberboard Sheathing high-density panels used for sheathing the exterior walls of stud walls

Fiberglass small-diameter glass strands that can be used loose, compressed into a mat, or woven into cloth for insulation, reinforcement, and the manufacture of a variety of products

Fiberglass Shingles a type of asphalt shingle that has a base mat of glass fibers, surfaced with mineral granules that provide weatherproofing and a fire-resisting barrier; a type of prepared roofing material (see also *Asphalt Shingles, Composition Shingles*)

Finger Joints light lumber created by cutting long points into the ends of each piece and then gluing them together

⚠ **Fire** the rapid, self-sustaining oxidation process with the release of heat and light; fire is the most costly natural hazard in the United States and causes about 4,000 fatalities annually; approximately 75 percent of all structure fires and 78 percent of all fire fatalities occur in residential construction

Fire Alarm an alarm system designed to warn building inhabitants of the presence of smoke, flames, or a rapid rise in heat

Fire-Alarm Control Panel a panel that contains the devices to control and test the fire-alarm system

Fire Area an area of a building enclosed by fire-rated construction capable of confining the spread of fire

Fire Assembly the assembly of a fire door, fire window, or fire damper, including all required hardware anchorage, frames, and sills; classes A, B, C, D, and E—from highest to lowest level of fire resistance—for protecting openings, separations, exit stairways, walls, and vertical shafts

Fire Barrier a fire-resistance–rated wall assembly of materials in which continuity is maintained, designed to restrict the spread of fire; may also refer to the separation of attached structures with a fire-resistive rating of not less than 1 hour and without openings (California Building Code, 2007)

Fire Blocking building materials installed to resist the free passage of flame to other areas of the building through concealed spaces; a material, barrier, or construction installed in concealed spaces to prevent the extension of fire for an unspecified period of time (Uniform Fire Code, 2007) **(Figure F-4)**

Figure F-4 Fire Blocking.

Fire Codes codes whose purpose is to control the spread of fire and allow sufficient time for the occupants of a burning building to exit safely, before the structure weakens to the extent that it becomes dangerous. Includes means of egress, the fire-resistance ratings of materials, and construction required for building and fire protection systems, depending upon size and use; a means of egress includes adequate exits access, exits and exit discharge; its construction, use, occupancy, and location limit a building's maximum size (height and area per floor); fire areas must be separated by fire-rated walls or barriers **(Figure F-5)**

Fire Cut the end of a wood joist or beam cut at a slight angle to allow the beam to fall away freely from a wall in the case of structural collapse; this self-releasing beam is designed to save the expensive masonry wall during a fire and resulting collapse

Fire Damper a listed device installed in ducts and air-transfer openings, designed to close automatically upon detection of heat to resist the passage of flames;

IBC (2000) Table 601		IA	IB	IIA	IIB	IIIA	IIIB	IV	VA	VB
NFPA 220 (1999)	I	I	II	II	II	III	III	IV	V	V
Table 7.2.2.2	443	332	222	111	000	211	200	2HH	111	000
BNBC (1999) Table 602	1A	1B	2A	2B	2C	3A	3B	4	5A	5B
SBC (1999)	I	II		IV	IV	V	V	III	VI	VI
Table 600				1Hr	U	1Hr	U		1Hr	U
UBC (1997)	I	II		II	II	III	III	IV	V	V
Table 6-A		FR		1Hr	N	1Hr	N		1Hr	N

Figure F-5 Comparison of the Construction Classifications of the Model Building Codes.

a dynamic fire damper is tested and rated for closure under elevated temperature airflow (California Fire Code, 2007); fire dampers are classified for use in either static systems, which will automatically shut down in the event of a fire, or in dynamic systems that operate in the event of a fire

Fire Department Connection inlet connections for fire apparatus to pump water into a standpipe, sprinkler, or combination system

Fire Division Wall a wall, extending continuously from the bottom to the top of a structure, that is designed and rated to retard the passage of fire in the structure

Fire Door a door designed to resist the passage of fire. Fire door assemblies consisting of a fire-resistive door, doorframe, and hardware are required to protect openings in fire-rated walls, so that the door area is no more vulnerable to fire than the wall; fire doors are rated by the amount of time they can resist the penetration of fire, typically from 45 minutes to 3 hours

Fire Endurance the length of time a material or an assembly can resist the passage of fire

Fire Escape a means of egress that is an alternate from the normal method and is isolated from the building proper to provide protection for people using it

Fire-Exit Hardware a door-latching assembly incorporating a device that releases the latch upon the application of a force in the direction of egress travel, which provides fire protection where used as part of fire door assembly

Fire-Fighting System part of a building's mechanical system that detects and extinguishes fires

Fire Load the measure of maximum heat release when all combustible material in a given fire area is burned; includes the content and structure of a building; the greater the fire load, the greater the possibility of structural failure during a fire; the potential fuel available to a fire is measured in pounds per square foot or BTUs per pound; ordinary combustibles such as wood and paper have approximately 8,000 BTUs per pound, whereas synthetics have approximately 12,000 BTUs per pound

Fire Partition an interior wall extending from one floor to the underside of the floor above; a vertical assembly of materials designed to restrict the spread of fire in which openings are protected (2007 California Building Code)

Fire Proofing any of various materials—such as concrete, gypsum, or mineral fiber—used in making a structural member or system resistant to damage or destruction by fire; a typical example would be covering structural steel members with masonry to protect them from the heat of a fire

Fire-Protection Systems building systems that may include heat and smoke detection, smoke control and heat-venting systems, fire alarm, and fire sprinkler and standpipe systems

Fire Rated noting or pertaining to a material, assembly, or construction as having a fire-resistance rating required by its use

Fire Rating a rating that indicates a material or system's ability to resist fire when tested by a recognized laboratory against applicable ASTM standards

Fire Resistance the ability of a material or assembly to resist combustion or burning in the presence of a fire, or the ability of a structural assembly to maintain its load-bearing ability under fire conditions

Fire Resistive Construction also known as Type I Construction, it is characterized by non-combustible, fire-resistive structural components **(Figure T-10)** (see also *Type I Construction*)

Fire-Resistance Rating the time in hours a material or assembly can be expected to withstand exposure to fire without collapsing, developing any openings that permit the passage of flames or hot gases, or exceeding a specified temperature on the side away from the fire. Rated fire resistance is a quality ascribed to a wall, floor, or column assembly that has been tested in a standard manner to determine the length of time it remains; ratings are provided after standardized testing by an independent testing organization and are used for typical wall, floor, and roof constructions, expressed in hours or portions thereof; NFPA 5000 requires calculations to establish rating of structural elements or assemblies to be performed in accordance with ASCE/SFPE 29, *Standard Calculation Methods for Structural Fire Protection;* NFPA 251, "Standard Methods of Tests of Fire Endurance of Building Construction and Materials," outlines the test procedures for fire-resistance ratings of structural elements and building assemblies; fire resistance-rated floor and roof assemblies shall be in accordance with NFPA 251, ASTM E119, or UOL 263

Fire-Resistant Construction refers to methods of controlling the spread of fire, increasing the length of exposure to fire a material can withstand without damage, and reducing a material's flammability; common materials used to provide fire-resistant protection include concrete (often with lightweight aggregate), gypsum concrete, gypsum board and plaster, and mineral fiber products (see also *Type I Construction*)

Fire-Resistant Panels panels created from Type X gypsum wallboard, which has natural fire resistance; fiberglass and other heat-resistant, noncombustible materials are added to the gypsum core to increase the natural fire resistance

Fire-Resistant Treated Plywood (FRTP) chemically treated plywood that resists fire, although some FRTP panels react and deteriorate in ordinary temperatures

Fire Retardant chemical or chemicals formulated to resist the spread of fire and/or increase the fire resistance of a material, which may be coated or treated with a fire retardant

Fire-Retardant–Treated Wood (FRT) any wood product impregnated with chemicals during manufacture that has been tested in accordance with ASTM E 84-05 or NFPA 255; not to be construed as "noncombustible," FRT pressure-treats wood, which decreases its allowed load by 25%

Fire Safety the measures taken to prevent fire or minimize the loss of life or property resulting from a fire, which includes limiting fire loads and hazards, confining the spread of fire with fire-resistant construction, the use of fire detection and extinguishing systems, the establishment of adequate firefighting services, and the training of building occupants in fire safety and evacuation procedures

Fire Separation any floor, wall, or roof–ceiling construction having the required fire-resistance rating to confine the spread of fire; also refers to the space required between a property line or adjacent building and an exterior wall having a specified fire-resistance rating

Fire Sprinkler Systems pipes located in or below ceilings, connected to a suitable water supply and supplied with valves or sprinkler heads made to open automatically at a certain temperature; the two major types of sprinkler systems are *wet-pipe systems* and *dry-pipe systems;* wet pipe systems contain water at sufficient pressure to provide an immediate, continuous discharge through sprinkler heads that open automatically in the event of a fire; dry-pipe systems are used when the piping is subject to freezing; pressurized air is released through the sprinkler head, allowing water to flow out

Fire Stop a specific system, device, or construction consisting of the materials that fill the openings around penetrating items such as cables, cable trays, conduits, ducts, pipes, and their means of support through the wall or floor openings to prevent the spread of fire **(Figure F-6)**

Fire Wall a fire-rated separation wall, usually extending from the foundation up to and through the roof of a building to limit the spread of fire; NFPA 5000 states that it shall be of not less than 2-hour fire-resistance construction in buildings of any type of construction; all openings are restricted to a

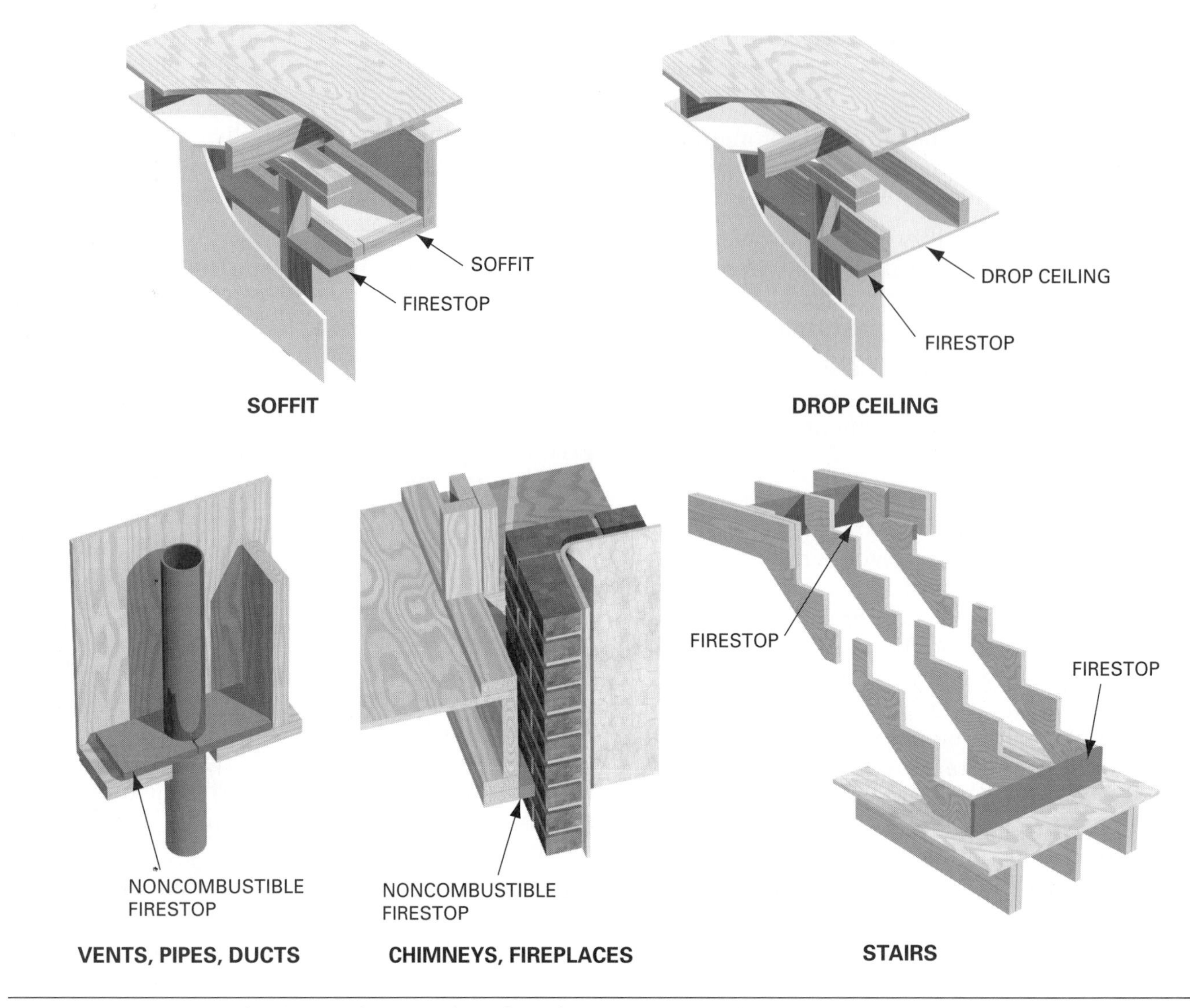

Figure F-6 Fire Stop. *(Courtesy of Western Wood Products Association.)*

certain percentage of the wall length and protected by a self-closing or automatic-closing fire assembly; fire walls are designed with sufficient structural stability under fire conditions to allow collapse of construction on either side with collapse of the wall. **(Figure F-7)**

Fire Window a window assembly rated in accordance with NFPS 257 and installed in accordance with NFPS 80

Fixed Beam a beam supported at two points and rigidly held in position at both points

Flameproof not burnable; noncombustible

Flame Retardant a compound used to raise the ignition point of a flammable material, thus making it more resistant to fire

Flame Spread the movement of flame across a combustible surface; a fire-resistance-rated wall assembly of materials designed to restrict the spread of fire in which continuity is maintained

Flame-Spread Rating a rating of how quickly a fire can spread along the surface on an interior finish material under test conditions as determined in accordance with ASTM Standard E84; the test is used to establish fire-resistant values in building materials, so that designers and builders know which materials are best for various applications; red oak flooring has a flame-spread rating of 100, whereas a cement asbestos board has a rating of 0; flame-spread rating has three classes: Class I ratings are from 0 to 25, Class II is from 26 to 75, and Class III is from 76 to 200; other flame-spread testing standards

Figure F-7 Fire Wall.

include NFPA 251, *Standard Methods of Fire Tests of Building Construction and Materials,* National Fire Protection Association; UL-263, *Fire Tests of Building Construction and Materials,* Underwriters Laboratories; and ASTM E119, *Methods of Fire Tests of Building Construction and Materials,* the American Society for Testing and Materials

Flammable material that can be easily ignited

Flammability the ability of a material to resist burning

Flange the top and bottom portion of an I-beam

Flashing material designed to prevent moisture from entering the wall or to redirect it; the exterior sheet metal used in roof and wall construction to keep water out, usually installed with exterior finishing material to prevent water leakage in places where it is likely to occur, such as at the intersection of a wall and roof or in the valley of a roof

Flash Point the temperature at which vapors given off by a fuel may be ignited but not sustain combustion

Flat Roof a roof with little or no pitch

Flat Slab Concrete Frame construction technique consisting of concrete slabs supported by concrete columns

Flat Slab Floor concrete slabs that are reinforced in two directions or more, usually isolated from the foundation by rigid insulation

Flex to bend or otherwise deflect without permanently deforming

Flight of Stairs a set of steps leading from one landing to the next

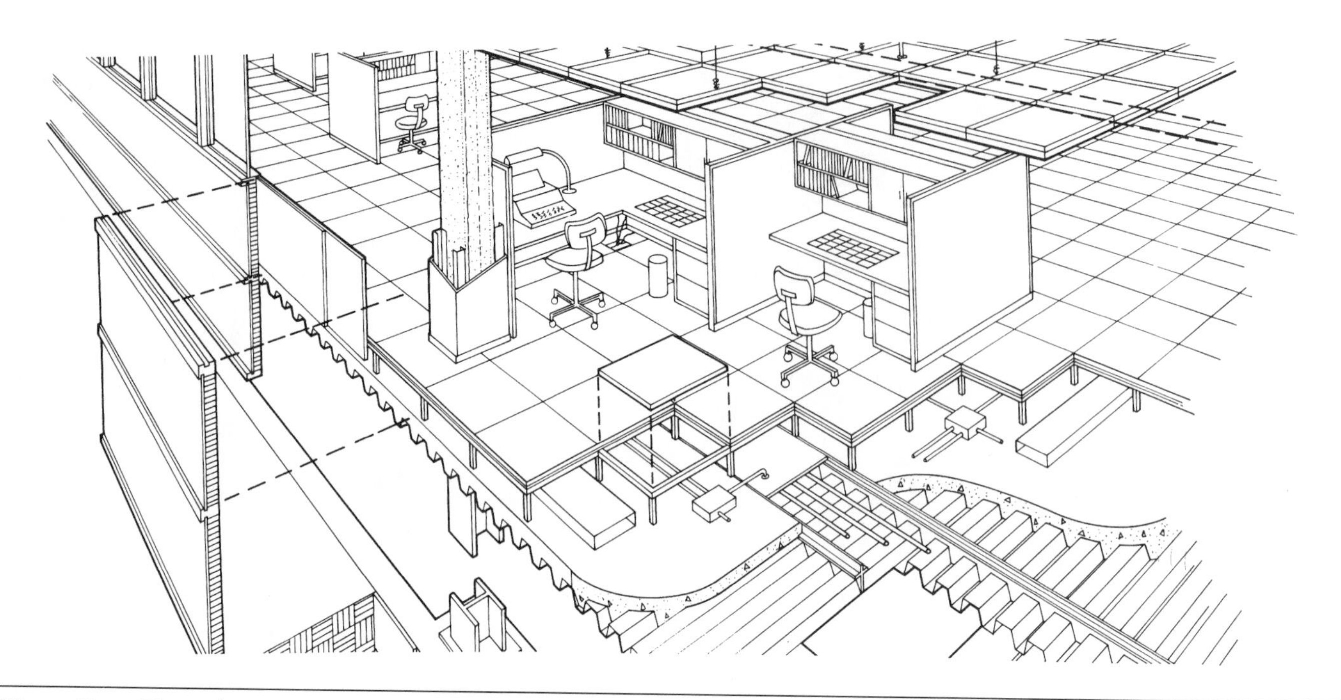

Figure F-8 Floor. *(Reproduced with permission from* The Building Systems Integration Handbook, Richard D. Rush, *Editor, Butterworth-Heinmann, 313 Washington St., Newton, Mass. 02158-1626.)*

F

Flitch Plate a metal plate bolted between two pieces of timber to form a girder

Floor the level, base, or horizontal surface of a structure that is designed to be walked on and to accommodate portable and permanent fixtures **(Figure F-8)**

Floor Failure examples of line-of-duty deaths due to floor failures include the following (causes included alteration deficiencies and overloading):

New York, NY 1966: Twelve firefighters died in a drug store
Boston, MA 1972: Nine firefighters died in a hotel
Brackenridge, PA 1991: Four firefighters died in a commercial building
Seattle, WA 1995: Four firefighters died in commercial building
Stockton, CA 1997: Two firefighters died in a single-family dwelling

Floor Framing the structural members of a floor that rest on the foundation or other structural members that support a vertical load

Floor Load measured in pounds per square foot to determine floor support systems; floor loads vary depending upon the type of structure and its use—for example, residential structures have a typical floor load of about 40 pounds per square foot

Flooring materials intended for use as a finished floor surface, such as wood planks, parquet, vinyl, and tile

Floor Panel an area of a floor between four supporting columns

Floor Plan a drawing of a building, looking down at the floor from above

Floor Systems a building's primary horizontal planes that support both live loads (occupants and contents) and dead load (all other fixed loads); structurally the floor system must transfer these loads laterally, to either beams or columns or to bearing walls, while providing lateral support for adjacent walls

Flue the chimney passageway for smoke and combustion produced in a fireplace or furnace

Fly Rafter a rafter at a gable end that is fastened to the roof sheathing above and the lookout blocks, which sit behind the rafter (see also *Barge Rafter, Barge Board*)

Flying Buttress an inclined bar of masonry carried on an arch transmitting the outward and downward thrust from a roof to a buttress separated from the wall and connected by an arch at the top

Foam Insulation a preformed, nonstructural, insulating board of foamed plastic or cellular glass. Rigid boards may be applied over a roof deck, over wall framing as sheathing, in cavity walls, or beneath an interior finish material; the foamed plastics are flammable and give off toxic fumes when burned; some rigid foam boards may be used in contact with the earth and are impervious to moisture

Foam Sheathing a rigid plastic foam insulating board used in or on the walls of a building to increase resistance to the transfer of heat or cold through the walls

Footing the base or bottom of a foundation, pier, wall, or column that is usually wider than the upper portion of the foundation; the footing spreads and transmits loads directly to the soil or pilings, and the added width at the bottom spreads the load over a wider area

Force energy that is exerted to perform work or cause a motion; forces include compression, tension, and shear forces, and force is also any influence that produces a change in the shape or movement of a body

Formwork a temporary structure made of wood, metal, or plastic and used to provide shape to concrete while it hardens; also known as *forms*

Foundation the base of a building, typically below the surface of the ground, designed to support and anchor the frame or wall system of the building and transmit its loads directly to the earth; the nature of the earth and the weight of the structure determines the foundation type, which usually includes concrete slabs and stem walls

Foundation Bolt a bolt set in wet concrete, used to bolt the mudsill to the concrete after the concrete has set

Foundation Mat a concrete pad reinforced by a steel rod that gives greater bearing area to concrete footings in a building

Foundation Vent openings through the foundation that permit ventilation under a structure; such vents are screened to prevent entry of small creatures

Foundation Wall a wall below the floor nearest grade that serves as a structural support for a wall, pier, column, or other structural part of a building, structure, or basement wall that resists lateral soil load

Foyer area or space within a building, located between the main entrance and the occupied areas of a building; when used in connection with a theater, the area between the lobby and the main floor

Fracture a break, a characteristic result of tension failure, usually resulting in actual separation of material

Frame the basic structural members of a building

Framework the skeleton or structural support

Framing typically refers to the wood or steel structural elements/members of a building as compared to a concrete or masonry wall; steel framing is often used in taller buildings and includes girders, columns, trusses, and beams; wood framing construction includes both balloon and platform framing

Framing Anchors metal fasteners used to connect framing members and to reinforce the joints; also known as *framing clips*

Free Standing Wall a wall that carries a compressive load and is not attached to another wall

Frieze a raised sculpture or ornamental decoration on a building; the horizontal part between the cornice and the wall

Furring wood or other material fastened to a surface prior to attaching gypsum wallboard or other wall covering panels; it can also be used to provide a level or plumb attachment plate for the panels

Furring Clip metal clips that are fastened at intervals to a wall that will be covered with masonry veneer

Fuse a safety device for an electrical circuit that melts and interrupts the circuit if the current exceeds a predetermined level for a specific amount of time

Fuse Box a distribution box for electrical circuits that contains fuses for the protection of the circuits

Fusible Metal a metal with a low melting point

Gable the triangular portion of wall enclosing the end of a pitched roof from ridge to eaves

Gable Dormer a dormer with a gabled roof

Gable End the portion of the end wall of a building between the top wall plate (eave to eave) and the point or ridge of the roof

Gabled End Vent a louvered ventilation device designed to remove heat from an attic space

Gabled Roof a roof that pitches in two directions

Gallery that part of the seating area of a theater or assembly room located above a balcony and having a seating capacity of more than ten (California Fire Code, 2007)

Galvanized chemically or electrostatically coated with zinc

Gambrel Roof a roof style in which the rafters are at two different slopes from the ridge to the eaves

Garden Apartments a general term for a number of different types of two- and three-story apartments, usually with separate apartments on each floor; these combustible multiple dwellings include row houses, townhouses, and similar structures

Geodesic Dome a structure composed of a formwork of triangularly arranged grid members whose finished appearance is a dome

Girder a large, horizontal structural member used to support beams or joists, typically refers to wood or steel elements that support the superstructure of the building; truss girders support smaller trusses

Girt a horizontal member spanning between exterior columns to support wall sheathing or cladding; also known as a *ribbon*

Glass Block glass, modular unit, usually 4 inches thick, mortared together for window panels, wall, and architectural features where light is desired

Glue-Laminated Beam/Timber a type of engineered wood structural member composed of many relatively short pieces of lumber glued and laminated together under pressure to form a long (25- to 100-foot), extremely strong beam; this composite, timberlike beam is able to withstand higher stresses, resist shrinkage and warpage, and has fewer wood defects. Also referred to as Glue Lams

Grade a reference plane representing the average of finished ground level adjoining the building at all exterior walls; also identifies the quality of lumber

Grade Beam a reinforced concrete beam supporting a bearing wall at or near ground level and transferring the load to isolated footing, piers, or pilings **(Figure G-1)**

Figure G-1 Grade Beam.

Gravel Roof a built-up roof with a final covering of gravel to protect the roof covering from the elements

Gravel Stop a raised metal edging installed around a roof that is to be covered with gravel-coated built-up roofing

Gravity the force that attracts all objects to the earth

Gravity Resistance System a term used in *Brannigan's Building Construction for the Fire Service* to describe all the structural elements and connections within a building that support and transfer gravity loads

Grillage a series of closely spaced beams designed to carry a particularly heavy load

Grout a thin, fluid concrete mortar for filling joints and cracks, often pumped to fill in cells of blocks

⚠ **Gusset Plate** a wood or metal fastener in the form of a flat plate, used to connect structural members such as trusses to stiffen and strengthen them; gusset plates in lightweight construction are stamped steel plates with teeth that project into the wood members **(Figure G-2)**

Figure G-2 Gusset Plate.

Gutter a trough along the eaves of a building that catches and redirects the flow of rainwater from the roof

Gypsum hydrous calcium sulfate, used as a core for drywall; this mineral substance is similar to plaster and has fire-resistant characteristics

Gypsum Board a noncombustible core, primarily of gypsum encased in paper, used for sheathing; a widely used interior finish material, it is also known as *gypsum wallboard, plasterboard,* and *drywall*; gypsum board is inexpensive, easy to install, and offers noise insulation and fire protection; fire resistance is reduced if joints are not taped; drywall panels are normally 4 feet wide and 8, 10, 12, 14 or 16 feet in length; there are six standard thicknesses available, ranging from 1/4-inch to 1-inch, and several types are available for a variety of applications; type X panels are used where fire-rated walls are required: For example, one 5/8-inch thick Type X gypsum drywall sheet installed on each side of a wood stud gives the wall a 1-hour fire rating

Gypsum Concrete a mix of gypsum aggregates and water

Gypsum Coreboard gypsum panels 1-inch thick, designed for use in solid gypsum partitions; the panels have a V-shaped tongue-and-groove edge

Gypsum Formboard gypsum panels designed to be used as the forms for poured gypsum concrete roof decks

Gypsum Lath a gypsum wallboard with either a solid or perforated face, which is designed to be used as the backing for plaster

Gypsum Planks gypsum roofing members, with steel reinforcement in the form of galvanized wire matting, cast into the planks 2 inches thick with tongue-and-groove edges

Gypsum Sheathing a water-repellent exterior gypsum wallboard used as a base for finish exterior building siding

Gypsum Wallboard same as gypsum board

H-Beam a structural member, usually of steel, with a cross section resembling the shape and proportions of the letter H

HVAC System see *Heating, Ventilation, and Air Conditioning*

Habitable Space a space in a building for living, sleeping, eating, or cooking

Half-Timber a type of building construction in which large wood members are left exposed on the outside walls, and the wall surface between them is covered with another material, such as stucco

Hardboard a high-density fiberboard typically in panels or sheets made from wood fibers compressed under heat and pressure to a density of 31 pounds per cubic foot or more

Hardwoods wood produced from broad leaf, deciduous trees such as oak, walnut, ash, maple, or birch

Head end of a block

Header a short, horizontal structural member in a wall spanning an opening such as a window, door, or stairway; also refers to a brick laid at right angles to the length of a wall, typically described as a *beam* in wood frame construction

Header Course a line of bricks with the ends of the bricks facing outward; also known as a *heading course* (see also *Bond Course*)

Head Jamb the upper horizontal finish member in a window frame

Head Joint the vertical joint between adjacent masonry units; also called a *cross joint*

Hearth a fireproof section of flooring extending out from a fireplace opening

Heartwood wood from the center of the tree to the sapwood, preferred for its strength and durability

Heating, Ventilating, and Air Conditioning (HVAC) controls the temperature, humidity, purity, distribution, and motion of the air in the interior spaces of a building; components include fans, ducts, plenums, filters, dampers, boilers, humidifiers, blowers, cooling towers, pipes, thermostats, registers, and diffusers; the Uniform Mechanical Code is the principal code regulating HVAC work **(Figure H-1)**

Figure H-1 HVAC System.

Heat Transfer the movement of heat from one place, fluid, or object to another

Heavy Floor one of the four construction classifications used by FEMA USAR teams to describe buildings for postfailure operations; others are *light frame, heavy wall,* and *precast concrete*; these buildings are frames of protected steel or cast-in-place, reinforced concrete with nonstructural exterior coverings—curtain walls, non–load-bearing concrete walls, and URM cover walls—that may be low-, mid-, or high-rise and of a variety of occupancies; the columns are aligned in a grid system up to forty feet apart, and they support roofs and floors, which may be prestressed, that have metal decks with lightweight concrete or concrete slabs and/or precast panels **(Figure H-2)**

Heavy Timber Construction also known as *Type IV construction*, characterized by noncombustible, fire-resistive exterior walls with large-dimensioned interior wood supports for floors and roofs; columns

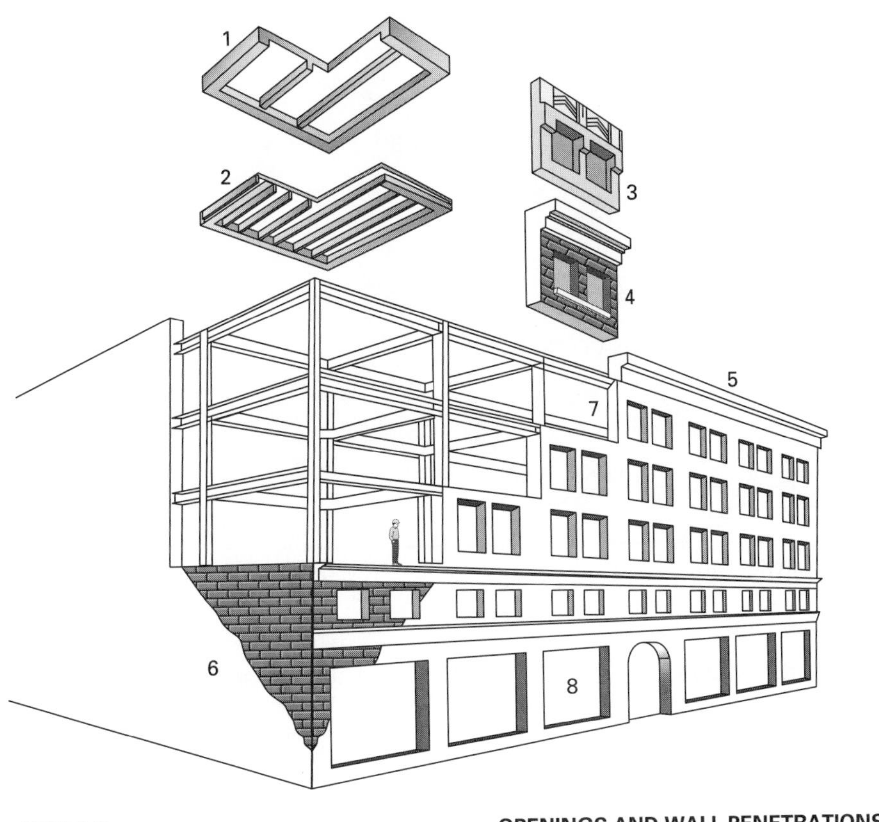

Figure H-2 Heavy Floor. *(Courtesy of FEMA/University of California at Berkeley.)*

are typically not less than 8 × 8 inches; beams and girders are not less than 6 × 10 inches; roof decking is not less than 2 inches (tongue-and-groove or splined planks or 11/8-inch wood structural panel); floor decking is not less than 3 inches (tongue-and-groove or splined planks with 1-inch tongue-and-groove flooring or half-inch wood structural panel subflooring); timber connectors include metal rings, plates, and bolts to transfer shear forces between the faces of two timber members; heavy timber construction can be recognized in the interior by the lack of void spaces **(Figure H-3)** (see also *Type IV Construction*)

Heavy Wall one of the four construction classifications used by FEMA USAR teams to describe buildings for postfailure operations; others are *light frame, heavy floor,* and *precast*; heavy-wall buildings are rigid, monolithic, and/or interdependent girder, column-and-beam structural systems typically used for commercial, mercantile, and industrial occupancies; they may be low-, mid-, or high-rise buildings; their load-bearing walls, compared to skeletal structural frames, may be of reinforced concrete, including tilt-ups, reinforced masonry, and unreinforced masonry (URM); floors and roofs may have steel decking with concrete topping, concrete slabs, wood diaphragms, and wood or steel trusses **(Figures H-4 a and b)**

⚠ **Hidden Voids** areas concealed from view that may allow fire and the products of combustion to travel **(Figure H-5)**

High-Rise any type of construction or occupancy having floors used for human occupancy located typically more than 75 feet above the lowest floor level

High-Slope Roofs roof slopes of four units vertical in 12 units horizontal (33-percent slope) or greater

Type of Element	Use	Size
Column	Supporting floor load	8- × 8-in. minimum any dimension
Column	Supporting roof load	6-in. smallest dimension, 8-in. depth minimum
Beams and girders	Supporting floor load	6-in. width and 10-in. depth minimum
Beams, girders, and roof framing	Supporting roof loads only	4-in. width minimum, 6-in. depth minimum
Framed or laminated arches	As designed	8-in. minimum dimension
Tongued and grooved planks	Floor systems	3-in. minimum thickness with additional 1-in. boards at right angles
Tongued and grooved planks	Roof decking	2-in. minimum thickness

Figure H-3 Heavy Timber Construction.

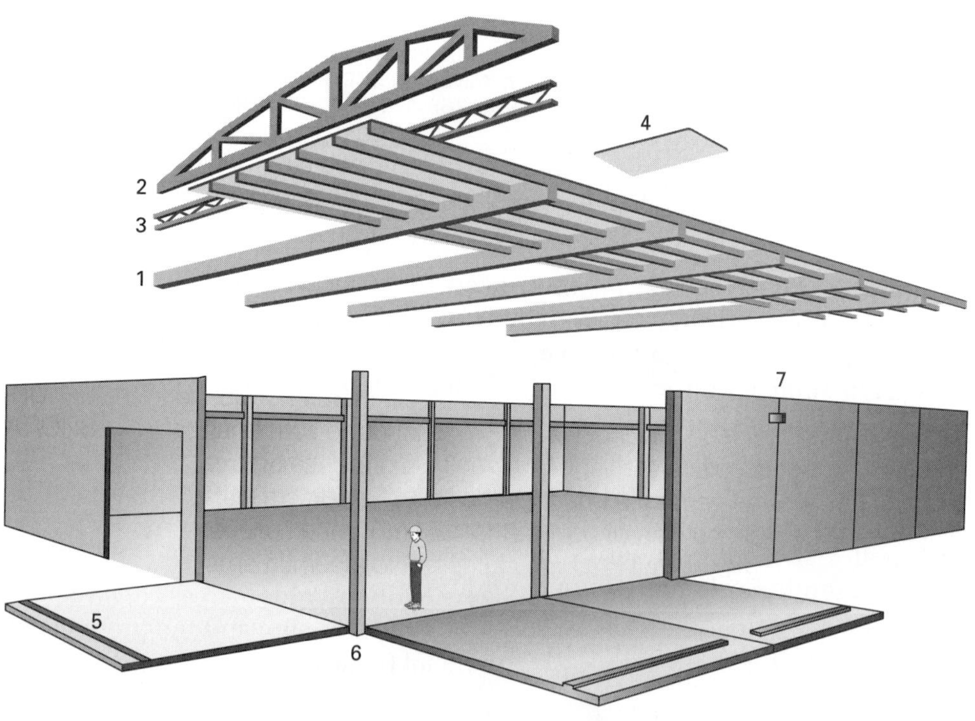

Figure H-4a Heavy Wall. *(Courtesy of FEMA/University of California at Berkeley.)*

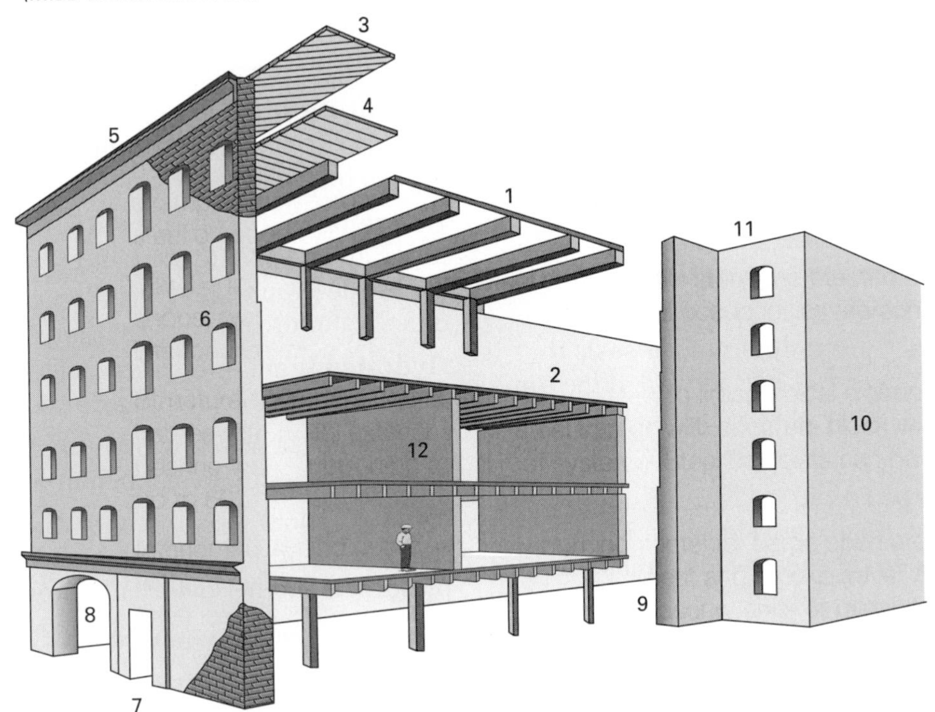

Figure H-4b Heavy Wall. *(Courtesy of FEMA.)*

Figure H-5 Hidden Voids.

Hip Roof a roof that slopes on the ends, as well as the sides, so that the eave line formed is constant on all walls; the sloping ends and sides meet at an inclined, projecting angle. Rafter types for hip roofs include hip rafters, jack rafters, hip jacks, valley rafters, valley jacks, and common rafters (see also *Roof* **(Figure R-09)**)

Hog Valley a roof valley between two roofs sloping toward each other, so that runoff water from the two roof slopes is directed toward a gutter

Hollow Concrete Masonry Unit a concrete masonry unit manufactured with one or more large holes throughout the score, showing an opening from side to side

Hollow Core a method of construction in which a framework, as for a door, is covered with thin plywood supported by pieces of lightweight material in the center of the framework

Hollow Masonry single-width construction using concrete masonry units with the voids on the ends of the blocks filled with grout as each course is laid

Hollow Masonry Unit a brick, tile, or block with one or more cavities

Hollow Tile a hollow masonry unit made of terra cotta

Hollow Wall a masonry wall of two parallel wythes with an air space between them but without ties to hold the wythes together

Homogenous composed of parts that are all of the same type of material

Hopper Window a window with a sash that is hinged at the bottom and swings inward at the top

Horizontal on a level plane with respect to the earth

Horizontal Diaphragm a rigid floor or roof structure acting as a flat, deep beam in transferring lateral forces to vertical shear walls, braced frames, or rigid frames

Horizontal Roof Area the area, taken in a flat plane, disregarding roof slopes that are covered by a roof

Horizontal Sliding Window a window with sashes that slide horizontally in tracks

Horizontal Thrust a force that tends to push in a horizontal direction

Humidity the moisture in the air

Hybrid Building a mix of building construction methods or materials that does not fit into one of the typical building types, I through V

Hydrant a standpipe with connections for fire hoses and a valve to start and stop the flow of water, set in the ground and supplied from a fire main or other water main

Hydrated Lime quicklime (calcium oxide) that has been mixed with water before packaging

Hydraulic a machine or system that is powered by applying pressure to a noncompressible fluid

Hydrocarbon a substance that is composed of hydrogen and carbon, such as kerosene, gasoline, plastics, and other petroleum products

Hydrous containing water

Hygroscopic a material that takes up and retains moisture

I-Beam a beam whose cross section is shaped like the letter *I* **(Figure I-1)**

Figure I-1 I Beams.

I-Joist a prefabricated and preengineered wood joist used for floor joists and roof rafters, and increasingly used in the place of dimensional lumber to frame floors, because they are generally lighter and more dimensionally stable than sawn lumber; i-joists are manufactured with sawn or laminated veneer lumber flanges along the top and bottom edges of a single plywood or OSB web; 10-inch joists span up to 16 feet, 16-inch joists span up to 25 feet, and deeper joists for commercial construction are available with spans up to 60'. **(Figures I-2 a and b)**

Impact Load a load that is in motion when it is applied, often considered a kinetic load delivered in such a short time that it can cause structural failure; examples include hose streams on a wall and firefighters jumping onto a roof

Impregnation the forcing of materials or chemicals into a substance; fire-retardant wood treatment uses impregnation with mineral salts that slow the rate of burning

Integrity the wholeness of a structure, especially with respect to the interaction of its elements

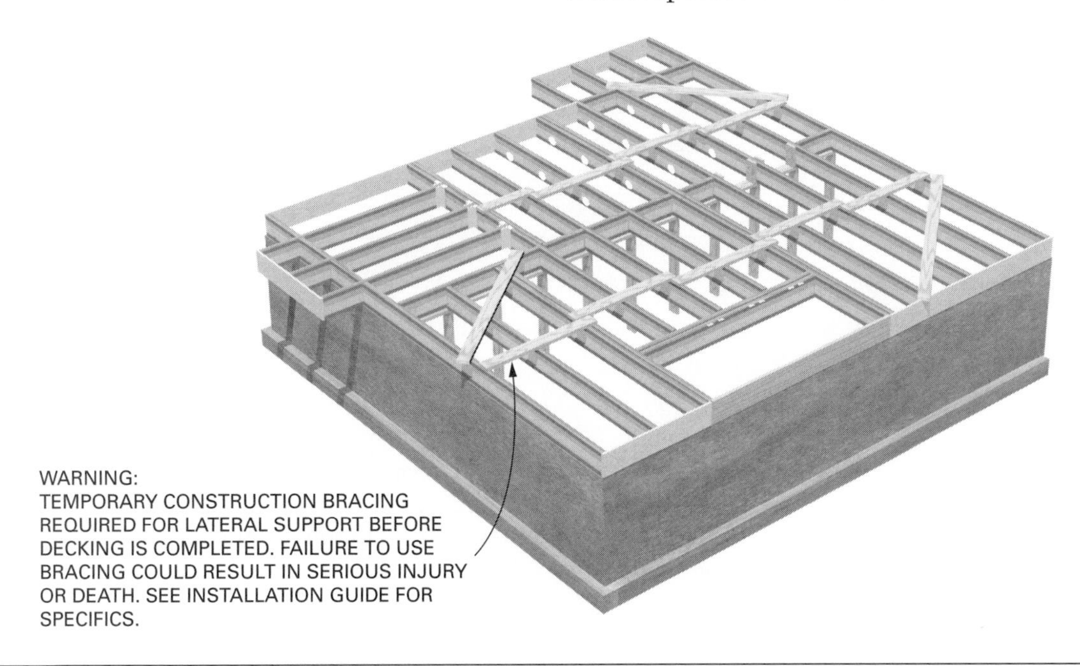

Figure I-2a I Joists. *(Courtesy of Louisiana Pacific Corporation.)*

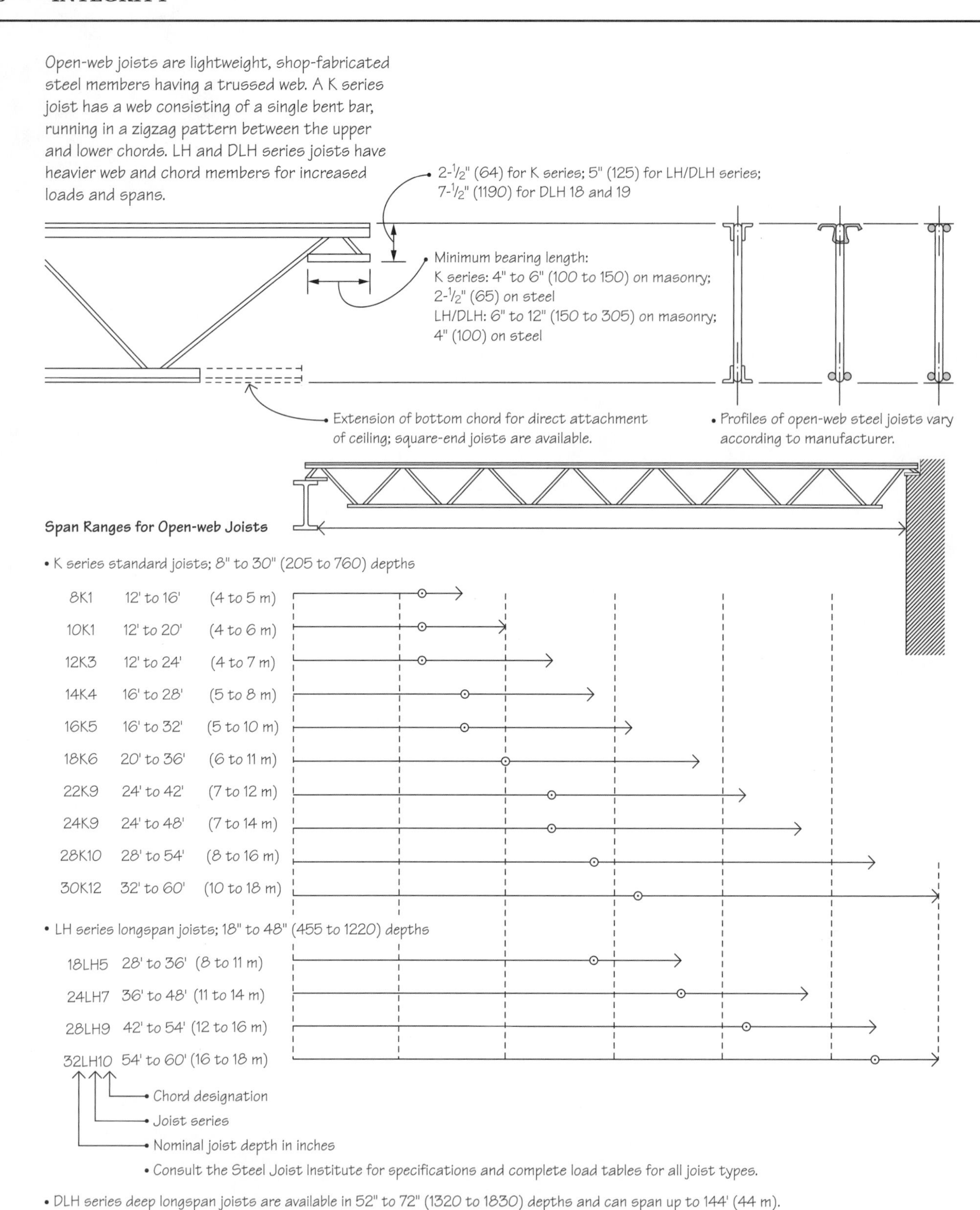

Figure I-2b I Joists. *(Courtesy of* Building Construction Illustrated, *Third Edition, by* Francis D.K. Ching, *© 2008, Reprinted with permission of John Wiley & Sons, Inc.)*

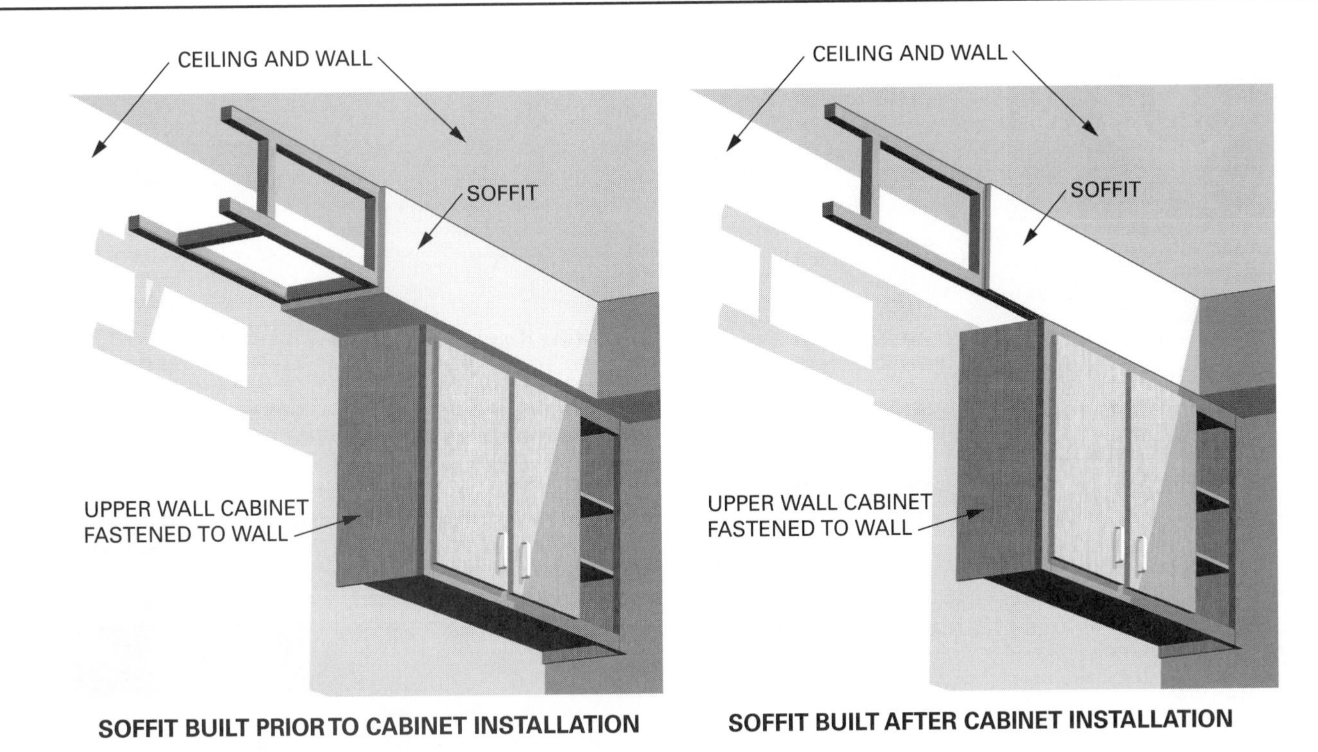

Figure I-3 Interior Soffits.

Interior Finish the exposed surfaces of walls, ceilings, and floors within buildings

Internal Skeleton sometimes referred to in high-rise buildings to describe the supporting columns and beams

Interlayment a layer of felt shingled between each course of a wood-shake roof covering

Interior Door a door not designed to be exposed to the elements, used to close a doorway on the inside of a building

Interior Soffits concealed spaces usually above cabinets, hallways, and bathrooms; expect and check for fire extension into concealed soffit spaces during kitchen fires **(Figure I-3)**

Interstitial Spaces inaccessible spaces between layers of building materials that create an intervening area for building equipment access and allow fire to spread unseen to other parts of a building (see also *Concealed Spaces, Voids*)

Iron a metallic element from which ferrous alloys such as cast iron are made; comes in various grades, types, shapes, and uses

I

J

Jack Rafter a roofing rafter that spans from a hip to a wall top plate or a valley to a ridge

Jamb the sides and tops of window and door frames

Joint a linear opening in or between adjacent assemblies, designed to allow independent movement of the building

Joist a horizontal structural framing member that supports the load of a floor or ceiling; joist types include light-gauge steel joists, conventional wood floor joists (typically 2 × 6 to 2 × 12 inches), prefabricated I-joists, prefabricated trusses (open-web, steel bar joists and parallel-chord wood floor joists), glue-laminated timbers, parallel strand lumber, and laminated veneer lumber (engineered wood products; **Figure J-1**) (See also *Wood Joist* **(Figure W-08)**)

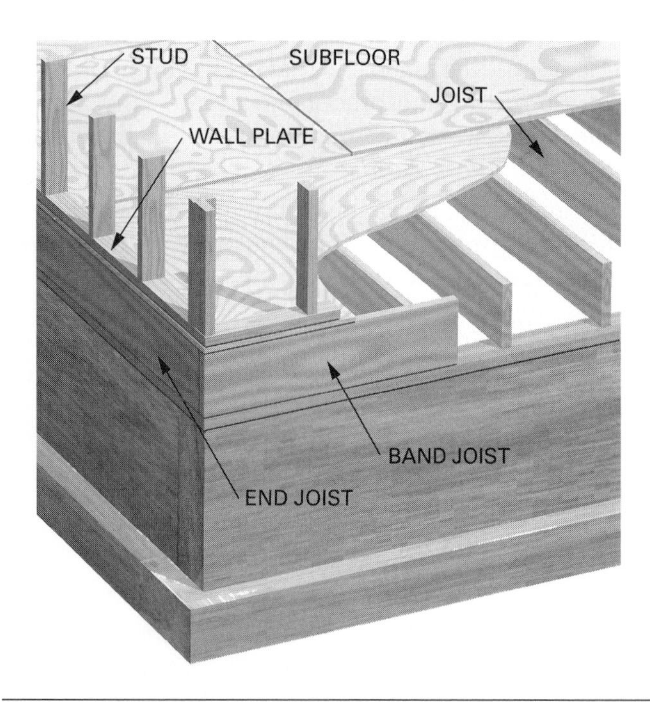

Figure J-1 Joist.

Joist Channel an opening in a masonry wall in which one end of a joist is supported

Joist Girder a large, open-web steel joist that support the top chord of smaller open-web steel bar joists; joist girders and steel beams often run parallel to exterior bearing walls

Joist Hanger a preformed sheet-metal fixture that is used to connect the end of a joist to a structural member at right angles to the joist **(Figure J-2)**

Figure J-2 Joist Hanger. *(Courtesy of Kathleen Siegel.)*

Joist Slab a type of floor system that uses a concrete slab with integrally cast T-beams or joists; also called a *ribbed slab*

Joist Tie-In a method of adding joists in which the new joist rests on the existing sill

Junction Box a metallic or nonmetallic box, designed with knockouts in the sides and back, used to support and protect electrical wire connections or conductor splices; several sizes and configurations of boxes are available for use with different wire sizes and quantities, but the National Electrical Code limits the number and sizes of wires that can be routed into each size of box

Junior Beam a smaller, secondary structural member used as a brace or to transfer the load between two other structural members

Keystone the wedge-shaped piece at the crown of an arch that locks the other pieces in place; the center stone at the top of masonry arch, it is called the *keystone* because it locks the other masonry pieces of the arch together, typically the last stone set in place (see also *Voussoirs*)

Keyway the groove cut in a member into which a square or rectangular key can fit for the purpose of locking two members together; in concrete construction, a groove or channel formed into one concrete pour is used to interlock another concrete structure poured at a different time, such as interlocking a wall to a footing

King Post the central vertical framing member of a pitched roof truss

King-Post Truss a style of roof truss that uses a central vertical framing member as the main link between upper and lower horizontal truss members; a pitched truss having a king post

Knee Brace a short diagonal member that provides lateral stability by joining columns and beams, used for reinforcing building framework, where an opening near the corner of the structure prevents the use of diagonal bracing; may also be a brace between a cantilevered member and the main structure to increase the load-carrying capacity of the member

Knee Wall short walls supporting rafters at some intermediate position along their length; may hide and extend fire **(Figure K-1)**

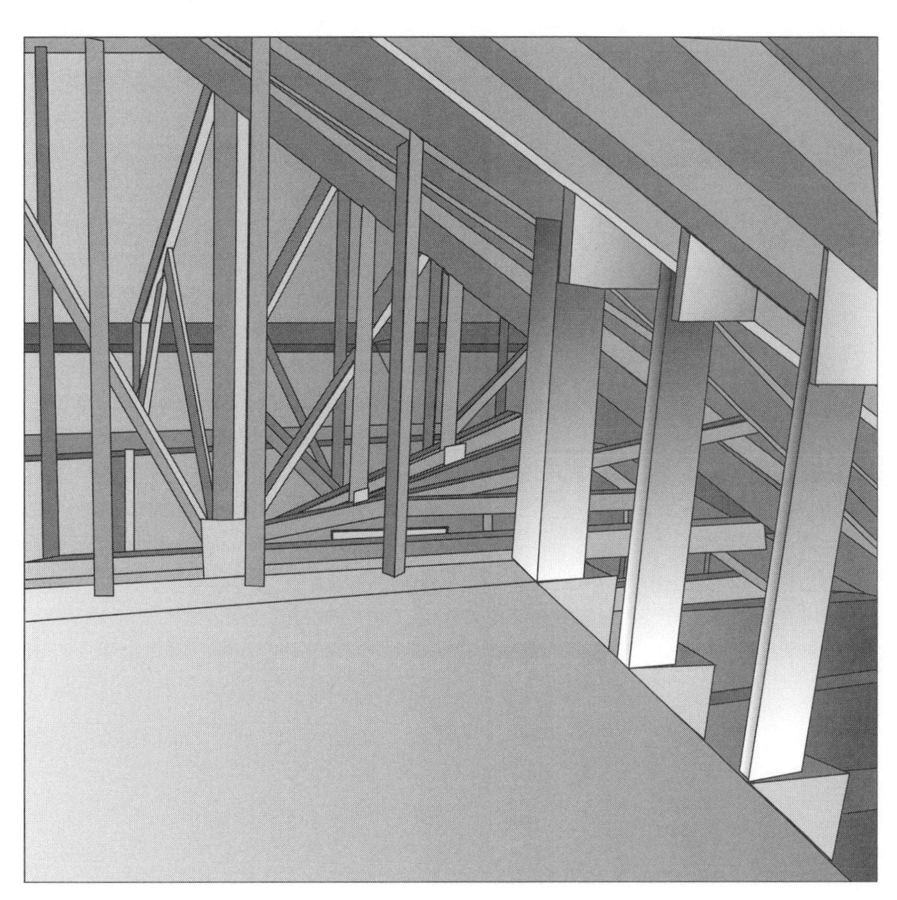

Figure K-1 Knee Wall.

Figure K-2 Knob and Tube Wiring.

Knob and Tube Wiring an obsolete form of house wiring in which the conductors are strung between porcelain standoffs (knobs), and porcelain tubes are used to line holes in structural members through which the wires pass **(Figure K-2)**

Lamella a thin, flat membrane or layer or short pieces of wood used in a roof framing method; the wood is laid out and joined in cross pieces to create a diamond pattern

Laminated in layers; two or more layers of material joined with adhesive, such as in glulam beams or plywood

Laminated Strand Lumber (LSL) lumber manufactured by bonding thin strands of wood, up to 12 inches long, with adhesive and pressure **(Figure L-1)**

PSL Column & Post Sizes	
Thickness	Width
3½″	3½″
3½″	5¼″
3½″	7″
5¼″	5¼″
5¼″	7″
7″	7″

PSL Beam Sizes	
Thickness	Depths
1¾″	7¼″
3½″	9¼″
5¼″	11¼″
7″	11½″
	11⅞″
	12″
	12½″
	14″
	16″
	18″

Figure L-1 Laminated Strand Lumber (LSL).

Laminated Veneer Lumber (LVL) a structural lumber product made by bonding layers of plywood with the grain all running in the same direction under heat and pressure using a waterproof adhesive; these engineered wood products are used as headers and beams or as flanges for prefabricated wood I-joists, and trade names include Microlam and Parallam

Lanai a covered porch, gallery, or veranda on the side or sides of a house

Landing a level platform between flights of stairs to break the climb or allow for a change of direction in the stairs; a platform at the top or bottom of a flight of stairs

Lateral acting to the side or at a 90-degree angle to an object

Lateral Force a force applied to the side of a structural member

Lateral Impact Load a force of short duration applied to the side of a member

Lateral Load a force applied to the side of a structural member

Lateral Reinforcement horizontal reinforcement surrounding a concrete column

Lateral Stability may be accomplished by a rigid frame, braced frame, or shear wall

Lateral Support a structural support designed to resist a load or force applied to the side of a structure

Lath closely spaced strips of wood used to fasten covering material to a wall or ceiling; in older buildings, wood lath was used to support plaster or stucco finish on the walls and ceilings; due to its combustibility, lath allows for more rapid fire spread in concealed wall cavities; today, gypsum and metal lath are the most commonly used types of lath, because they are inexpensive and can be installed in sheets, which cover a large area very quickly **(Figure L-2)**

Lean-To a small building with a shallow roof that slopes in only one direction

Lean-To Rafter a rafter that extends from the top plate of the high wall to the top plate of the low wall on a lean-to roof

Figure L-2 Metal Lath.

Lean-To Roof a roof that has only one slope, similar to half of a gable roof; the slope of this roof is often fairly shallow

Ledge a narrow, projecting shelf

Ledger a horizontal support member fastened to a wall or a beam on which joists rest; also known as a *ledger board* or *ledger strip, ribbon* or *ribbon strip* (see also *Ribbon* **(Figure L-3)**)

Figure L-3 Ledger.

Length a linear dimension along the longest side of an object or structure

Let-In to insert into the surface of a stud, wall, or the like as a permanent addition; an example would be a section of stud being notched, so that a let-in diagonal brace surface fits flush with the stud surface

Lift-Slab Construction a technique used in the construction of multistory concrete structures; concrete floor and roof slabs are cast at the ground level, posttensioned, and lifted into place using columns and hydraulic jacks **(Figure L-4)**

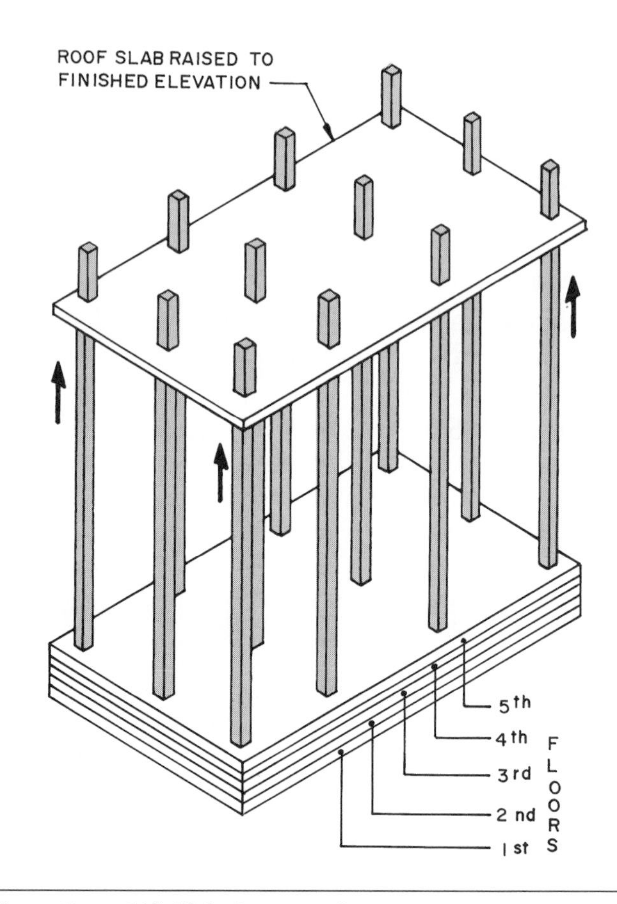

Figure L-4 Lift Slab Construction.

Lightweight Steel Framing structural steel framing members made from cold-rolled lightweight sheet steel

Light Framing lumber, up to 4 inches wide, used to construct the skeleton of a structure such as a house, mobile home, or other one- or two-story building; it can also be used to build portions of taller structures

Light Frame one of the four construction classifications used by FEMA USAR teams to describe buildings for postfailure operations; others are *heavy wall, heavy floor,* and *precast concrete*; light frame structures include wood and light-gauge steel skeletal structural frames; light wood frame buildings are typically low-rise residential and multiple-occupancy structures, typically one to three stories; light-gauge steel frame buildings are commercial, light manufacturing, and multiple-residential occupancies, typically one to four stories, and occasionally up to six stories **(Figure L-5)**

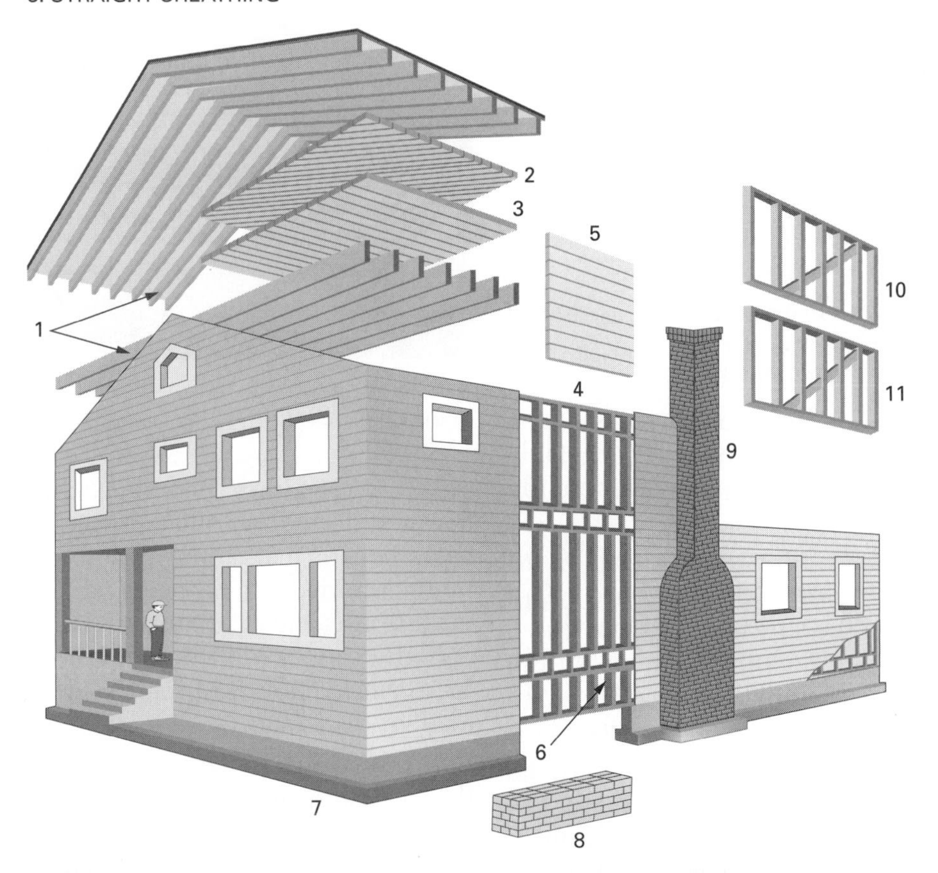

Figure L-5 Light Frame. *(Courtesy of FEMA/University of California at Berkeley.)*

Light Frame Construction a common contemporary construction method for Type V combustible wood framing, it replaced the use of heavy timber wood framing; recent improvements resist lateral dynamic loads; a popular system for light-framed structures is metal-plate-connected wood roof trusses based on standard designs; care must be taken during construction not to damage critical truss members and connections during transport, erection, and installation; proper temporary bracing is important to avoid progressive collapse **(Figure L-6)** (see also *Type V Construction* **(Figure T-14)**)

Light Framed Wall wood or steel studs are used to construct a wall, compared to a skeletal frame of steel or concrete columns

Figure L-6 Light Frame Construction.

L

Lightweight Concrete concrete that weights 90 to 110 pounds per cubic foot because of the use of lightweight aggregate; regular concrete weighs approximately 150 pounds per cubic foot

Lime a caustic form of calcium oxide used in cement, mortar, and plaster; used with sand as mortar until around 1880, when Portland cement became widely used

Limited Combustible Material refers to building construction materials that do not comply with the definition of noncombustible material

Limiting Fire Spread four generally accepted construction methods to limit fire spread may be found in the California Fire Code (2007):

Interior finishes The interior surfaces of the building shall not contribute to an unacceptable rate and magnitude of fire spread and generation of heat and smoke.

Concealed spaces The construction of concealed spaces shall not contribute to an unacceptable rate of the spread of fire, hot gases, and smoke to areas of the building remote from the fire source and shall limit their spread beyond the immediate area of the origin of the fire.

Compartmentation The building shall be compartmented by walls and floors, including their associated openings with proper closures, to limit the spread of fire, hot gases, and smoke to an acceptable area beyond the immediate area of fire origin.

Structural integrity The building's structural members and assemblies shall be provided with the appropriate degree of fire resistance to limit structural damage to an acceptable level and to limit damage to the building and its contents and to adjacent buildings and property.

Linoleum a tile or sheet floor covering of ground cork, wood filler, and pigment, held together by linseed oil or other binders and backed with felt

Lintel a member above a door or window opening that supports the load above it and allows the compressive stresses to flow around the opening to adjacent sections of the wall; may be made of wood, concrete, masonry, reinforced brick, steel angle, or precast concrete (**Figure L-7**)

Figure L-7 Lintel.

Listed equipment or materials that comply with approved codes, standards, or tests and have been found suitable for use in a specified manner

Live Load the load on a structure, usually described in pounds per square foot, caused by temporary or movable sources that are not part of the building: furniture, people, or other movable loads not included as a permanent part of the structure; the basic capacity of a structure to carry a live load is predetermined and becomes part of the requirements for building the structure; live loads do not include construction or environmental loads such as wind load, snow load, rain load, earthquake load, flood load, or dead load.

Load a weight or force on something; any effect that a structure must be designed to resist; forces or other actions that result from the weight of all building materials, occupants and their possessions, environmental effects, differential movement, and restrained dimensional changes.

Loading refers to the weight of materials or objects in a building and is usually described in pounds per square foot (psf). A building must be able to support two types of loads, *static* and *dynamic*.

Static and dynamic loads Static loads remain constant and are assumed to be applied slowly until a peak value is reached. Static loads include live loads, dead loads, impact loads, snow loads, rain loads, and occupancy loads. Dynamic loads are applied suddenly to a structure the two major types of dynamic loads are wind loads and earthquake loads

Imposition of loads refers to the manner in which loads are imposed on the structure

axial loads pass through the center

eccentric loads are an off-center force

torsional loads are forces that cause twisting of the supporting members. Architects and engineers consider a building's occupancy when determining uniform and concentrated floor live loads (e.g., 40 psf for residential, 50 psf for office, and 75 psf for retail) as well as for roof dead and live loads.

Other frequently used load terms include the following:

Concentrated load a load applied at one point or within a limited area of a structure

Dead load the weight of the building and any equipment permanently attached to it Concentrated and dead loads are static loads that include walls, floors, columns, girders, HVAC units, and fire escapes

Lateral load a load applied to a structure from a direction parallel to the ground

Live load a movable load, such as people and water

Wind load a lateral load imposed on a structure by wind

Occupant load the number of occupants a room or building may safely hold

Load duration the length of time over which a load is applied

Impact load load that is in motion when it is applied (explosion, master stream, moving vehicles)

Fire load the amount of combustibles measured in BTUs per square foot

Permanent loads are those loads in which variations over time are rare or of small magnitude; other loads are considered to be *variable loads*

Load-Bearing Partition an interior wall that a carries part of the load of the structure

Load-Bearing Element any column, girder, beam, joist, truss, rafter, wall, floor, or roof sheathing that supports any vertical load, in addition to its own weight, or any lateral load (California Fire Code, 2007)

Load-Bearing Wall a wall defined by columns and beams that transmit compressive forces to the ground; load-bearing walls support other walls, floors, or roofs, compared to non–load-bearing walls that support only their own weight

Load Capacity the maximum weight a material can support before collapse occurs

Load Effects forces and deformations produced in structural members by the applied loads

Load Factor a factor that accounts for deviations of the actual load from the nominal load and for the probability that more than one extreme load will occur simultaneously (California Fire Code, 2007)

Loading the weight of building materials or objects in a building

⚠ **Load Paths** the concept that loads must be continuously carried to the ground; the structural frame of a building needs to withstand all possible loading conditions and forces of compression, tension, and shear; a uniformly distributed load is measured in pounds per square foot (psf); architects and engineers doing structural design select members that safely and economically carry anticipated loads; gravity load paths may be visualized in a "top-down" approach, considering that a structure's roof and floor system and beams are connected to columns and/or bearing walls; lateral load paths may be viewed as a "pushover" or from the side; the concept includes connections with multiple load-resisting systems (redundancy) and ductility; the CFC (2007) states that "any method of construction to be used shall result in a system that provides an identified, continuous load path capable of transferring all loads from their point of origin through the load-resisting elements to the ultimate point of support"; a force may cause a building element to fail due a change or shift in the load path; the failure of an element occurs when the load exceeds the limit or capability of that element **(Figure L-8)**

Figure L-8 Load Path.

Load Stress an internal stress created by a load in a structural element, including compression, tension, and shear stresses; a collapse (failure) can occur when a stress created by a load exceeds the load-carrying capability of the structural element

Figure L-9 Load Transfer.

Figure L-10 Lookout Rafters.

Load Transfer reactions redistributed among various supports due to differential settlement; load transfer may result in significant stress increases in certain supports **(Figure L-9)**

Lookout Rafters rafters that support the roof overhang **(Figure L-10)**

Louvers slats placed at a downward slanting angle, in close proximity but not touching, so there are openings between them; they are used primarily for ventilation, allowing the passage of air and light but preventing rain from entering

Low-Density Concrete concrete combined with a lightweight aggregate, such as expanded vermiculite or perlite, or foaming agents, which create air pockets, resulting in a mixture that weighs 50 pounds or less per cubic foot, instead of 150 pounds per cubic foot for normal concrete; low-density concrete is used where weight would be a detriment, such as for suspended concrete floors

Low-Rise a building one to three stories in height

Low-Sloped Roof from 2.5 units vertical to 12 units horizontal up to four units vertical to 12 units horizontal (21- to 33-percent slope)

Lumber sawn and sized lengths of wood used for building; lumber includes boards and structural lumbers; boards are less than 2 inches thick and are used for siding, subflooring, and interior trim; structural lumber is wood graded on the basis of strength and intended use and includes dimension lumber and timbers; dimension lumber ranges from 2 to 4 inches thick and 2 or more inches wide and is used for joists, planks, light framing, and decking; timbers are 5 or more inches thick and include beams, stringers, posts, and timbers; *Nominal dimensions* are the dimensions of a piece of lumber before drying and surfacing, and *dressed sizes* are the actual dimensions of a piece of lumber after seasoning and surfacing; lumber is generally available in lengths from 6 to 24 feet in multiples of 2 feet; parallel strand lumber and laminated veneer lumber are recent structural engineered lumber products that are manufactured, compared to solid sawn lumber

Main Runner the longitudinal metal support that runs the length of a room and holds the ceiling tiles of a suspended ceiling; main runners are supported from the structure above them by a length of wire fastened to the existing ceiling, ceiling joists, or overhead beams

Mall a roofed or covered common pedestrian area within a mall building that serves as access for two or more tenants and does not exceed three levels that are open to each other

Mansard Roof a type of roof with two different slopes around all sides of the structure, the upper of which may be nearly horizontal, the lower nearly vertical, and a roof having on each side a steeper lower part and shallower upper part; *False mansards* are common on many new commercial buildings and may be added to older buildings to improve their appearance; usually made out of wood or steel, they provide a horizontal attic space and present a collapse hazard as well as an avenue for fire and the products of combustion to spread **(Figure M-1)** (see also *False Mansard* **(Figure F-02)**)

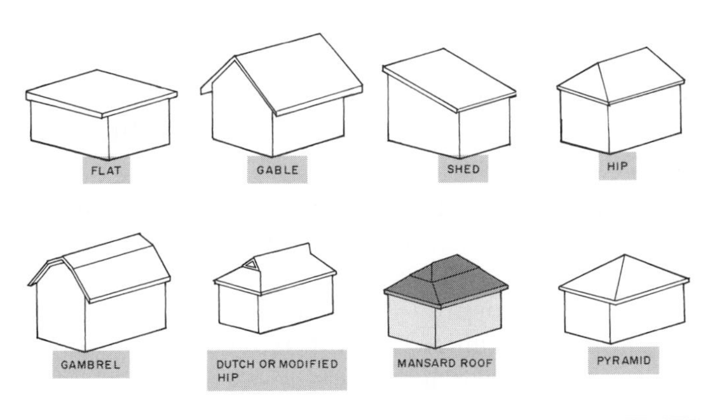

Figure M-1 Roof Examples.

Mantel a shelf over a fireplace opening

Marquee a permanent canopy or projection attached to and supported by the building that projects into the public right-of-way, typically seen at the entrances of theaters and hotels

Masonite trade name for a dense, compressed board made from wood products and a binder, used as a nonstructural building material, such as an underlayment for vinyl flooring or similar application, where a smooth, flat surface is needed

Masonry the use of bricks, stones, tile, concrete blocks, or other units set in mortar for construction purposes, with qualities that include inherent compression strength, thermal and fire-resistive properties, and ease of use

Masonry Anchor metal braces embedded into the masonry, used to attach masonry to other structures or to attach additional members, such as doorjambs and window frames, to the masonry; masonry anchors include anchor bolts, Z-type rigid anchors, and dovetail corrugated brick-tie and expansion sleeves

Masonry Block a hollow concrete building block; also referred to as a *concrete masonry unit,* or *CMU* (see also *Concrete Block*)

Masonry Bond Beam a specially designed concrete block that is hollow on top, used as one or more courses in the building of a concrete block wall; rebar is placed in the hollow of the blocks horizontally along the length of wall, and the blocks are filled with concrete; the concrete beams add reinforcement around the perimeter of a building

Masonry Cement Portland cement with air-entraining additives and inert limestone or hydrated lime added; these materials improve the workability, plasticity, and water retention of the cement

Masonry Columns masonry bracing incorporated into masonry walls to stabilize walls and support concentrated loads; masonry columns have vertical reinforcement in a grout-filled core; pilasters and buttresses provide the same function but in a different manner

Masonry Construction the modern construction use of masonry materials that include brick, clay tile block, concrete block, and stone material in conjunction with mortar, grout, and reinforcing steel

Masonry Joint the joint between masonry units; various terms are used to describe the mortar joint cross-section profiles, such as *struck joint, weathered joint, raked joint, flush joint, concave joint,* and *V-joint*

Masonry Lintel a header for doors and windows made of precast concrete or block, mortar, and rebar; they are used in masonry walls so that the lintel material is compatible with the appearance and durability of the wall

Masonry Reinforced Wall a wall constructed of stone, clay, cement, bricks, or concrete that has been reinforced with steel

Masonry Veneer a masonry facing for a wall, as opposed to a wall in which the masonry serves as the structural support; an example of masonry veneer would be a single layer of brick applied to the exterior of a wood-framed house

Masonry Walls refers to brick, concrete block, structural clay tile, structural glass block, or natural or cast stone walls that may be constructed as solid walls, cavity walls, or veneered walls; applications include bearing walls, shear walls, curtain walls and building facades; perimeter masonry walls are often bearing walls, and masonry walls may be unreinforced or reinforced; older masonry walls (pre-1940) were often unreinforced masonry, or URM (lacking steel reinforcement); URM construction used lime-cement-mortar joints, which tend to weather and weaken; problems with unreinforced masonry construction have been a lack of tension strength and ductility, which make these among the most hazardous forms of building construction found in the United States; masonry walls are not designed to resist lateral impact loads; masonry buttresses, columns, arches, and bearing walls utilize the compressive strength of masonry and recognize its weakness in tension; masonry bearing walls are typically arranged in parallel sets to support steel, wood, or concrete spanning systems; common spanning elements include open-web steel joists, timber or steel beams, and concrete slabs; pilasters stiffen masonry walls against lateral forces and provide support for large concentrated loads; openings in masonry walls may be arched or spanned with lintels; masonry wall thickness is a minimum of 6 inches, for reinforced masonry bearing walls and solid masonry walls in one-story buildings not more that 9 feet high, and 8 inches for masonry bearing walls and shear walls

Masonry Wall Ties metal devices made of mesh, corrugated metal, or heavy wire that are embedded in mortar between masonry units and connected to an existing wall

Mass the amount of material found in a given member

Materials those items used to construct an object or structure; the primary materials used in building construction today in the United States include wood, steel, concrete, and masonry; these materials may be found together or separately within the same building or structural element; cost, quality, engineering capabilities, and fire-resistant capabilities all determine a material's use in a building **(Figure M-2)**

Means of Egress a continuous and unobstructed path of vertical and horizontal egress from any occupied portion of a building or structure to a public way; it consists of the *exit, exit access,* and *exit discharge*

Means of Escape a way out of a building or structure that does not conform to the strict definition of means of egress but does provide an alternate way out

Members primary, secondary, or tertiary structural members; primary or main members are essential to the stability of a structural whole; a secondary member is supported by a primary member, and a secondary member supports tertiary member **(Figure M-3)**

Membrane a thin sheet separating two materials or surfaces, typically referred to as a *thin coating* or *sheet* of flexible material, usually water-resistant, such as the elastomeric sheets used to waterproof roof surfaces

Membrane Ceiling usually refers to a suspended, insulating ceiling-tile system

Membrane Fireproofing a ceiling of noncombustible tiles or other material protecting steel structural members

Membrane-Penetration Fire Stop a material, device, or construction installed to resist, for a prescribed time period, the passage of flame and heat through openings in a protective membrane to accommodate cables, cable trays, conduit, tubing, pipes, or similar items

Membrane Roof a single layer of waterproof synthetic membrane over insulation on a roof deck (see also *Single-Ply Roof*)

Membrane Structures thin, flexible surfaces that carry loads primarily through the development of tensile stresses; they may be suspended or stretched between posts or be supported by air pressure

Metal a class of chemical element having a crystalline structure and the ability to conduct heat and electricity

Performance of Common Building Materials under Stress and Fire

Material	Compression	Tension	Shear	Fire Exposure
Brick	Good	Poor	Poor	Fractures, spalls, crumbles
Masonry block	Good	Poor	Poor	Fractures, spalls
Concrete	Good	Poor	Poor	Spalls
Reinforced concrete	Good	Fair	Fair	Spalls
Stone	Good	Poor	Fair	Fractures, spalls
Wood	Good w/grain; poor across grain	Marginal	Poor	Burns, loss of material
Structural steel	Good	Good	Good	Softens, bends, loses, strength
Cast iron*	Good	Poor	Poor	Fractures

*Some cast iron may be ornamental in nature and not part of the structure or load bearing.

Figure M-2 Materials.

Figure M-3 Members.

Metal Cladding panels typically 3 feet wide that span vertically between horizontal steel girders spaced 8 to 24 feet apart, used primarily to clad industrial-type buildings

Metal Composite Material (MCM) a factory-manufactured panel consisting of metal skins bonded to both faces of a plastic core

Metal Decking a ribbed, corrugated steel sheet used for formwork for site-cast concrete floors and roofs; Typically the panels are welded to the steel joists or beams and fastened to other panels with screws, welds, or seams; three major types of metal decking are *form decking, composite decking,* and *cellular decking,* but also referred to as *Q-deck* or *Robertson decking* **(Figure M-4)**

Figure M-4 Metal Decking.

Metal Drip Edge a preformed piece of sheet metal placed along the edge of a roof to encourage water to run off the edge of the roof rather than flow back under the shingles or eaves

Metal Form a metal pan or other shape used to mold and support fluid concrete

Metal Lath thin metal sheets that have been expanded or stretched and then stamped through with a slotted pattern, giving them a screenlike appearance; they are used as backing and structural support for plaster, stucco, ceramic tile, and mortar **(Figure M-5)** (see also *Lath* **(Figure L-2)**)

Metal Roof Panel an interlocking metal sheet having a minimum installed weather exposure of 3 square feet per sheet

Metal Roof Shingle an interlocking metal sheet having an installed weather exposure less than 3 square feet per sheet

Metal Roofing metal shingles or sheets that typically have interlocking seams, ridges, edges, or a corrugated shape

Metal Roof Decking typically corrugated steel panels that may be covered with insulating panels, concrete, and roof covering; common on steel framed and masonry buildings (see also *Metal Decking*)

Metal Siding any metal in sheets or strips used on the exterior of a building

Metal Stud a preformed steel framing member that is protected against corrosion by galvanizing or coating it with aluminum or an aluminum–zinc compound; studs come in different thickness for load-bearing or non-load bearing applications, but metal studs are most commonly used in commercial buildings and for tenant improvement (T&I) in the creation of partition walls

Mezzanine an intermediate level between the floor and the ceiling, which projects like a balcony between two floors, having less than a third of the total floor area of any other level in the building

Microlam a brand name of a laminated veneer lumber used for headers and beams or as flanges for prefabricated wood I-joists

Mid-Rise a building four to seven stories in height

Mild Steel steel with relatively low carbon content and without special alloying elements, used in the production of nails, structural beams, brackets, and other building hardware

Miscible capable of being mixed

Mobile Home a dwelling unit on a chassis with axles and wheels, which can be moved and mounted on a temporary foundation or anchored to a semipermanent foundation

Modification the alteration, change, or reconfiguration of any space; examples include changing the roofline of a building, adding or eliminating a door or window, reconfiguring or extending a system, or installing additional equipment

Modified Bitumen Roof Covering one or more layers of polymer-modified asphalt sheets, which are fully adhered or mechanically attached to the substrate or held in place with an approved ballast layer

Modular Homes houses that are manufactured then factory-assembled into the largest practical sections or modules that can be transported to a building site

Molding or Moulding a length of material that has been milled into a pattern, used to cover gaps, such as between the wall and floor, or to provide a finished or decorative appearance

Moment an engineering term that describes a force that acts at a distance from a point and that tends to cause the body to rotate about that point

Monolithic consisting of one piece of stone or stonelike material, such as concrete

Monolithic Concrete Construction a continuous concrete pour and bond between all structural elements, except for construction joints

Mortar a mixture of Portland cement, fine aggregate or sand, lime, and water in proportions that will yield a plastic mixture that can be used to fill voids between masonry units such as brick, block, or stone; mortar acts as a bonding agent to hold units together and add strength to the structure as a whole; mortars are classified by the characteristics obtained in the mixture: type M mortar is a durable mortar with high (2500 psi) compressive strength that is used for reinforced brick masonry, foundations, retaining walls, sewers, and catch basins; type N mortar is a medium strength, waterproof mortar used in areas above grade with high exposure to the elements, areas such as chimneys and exterior walls; type O mortar has medium-low compressive strength and is recommended for solid masonry and nonbearing interior walls; type S mortar has medium-high (1800 psi) compressive strength with high tensile strength and is used for walkways, stucco, or load-bearing applications at or below grade; mortar types N, O, and K have lesser compressive strengths and are used for non–load-bearing walls or in exposed or decorative masonry areas

Mortar Joint masonry joint bonded with mortar

Mortise a recess or slot cut into a board that receives the projecting portion (tenon) of another member to form a joint

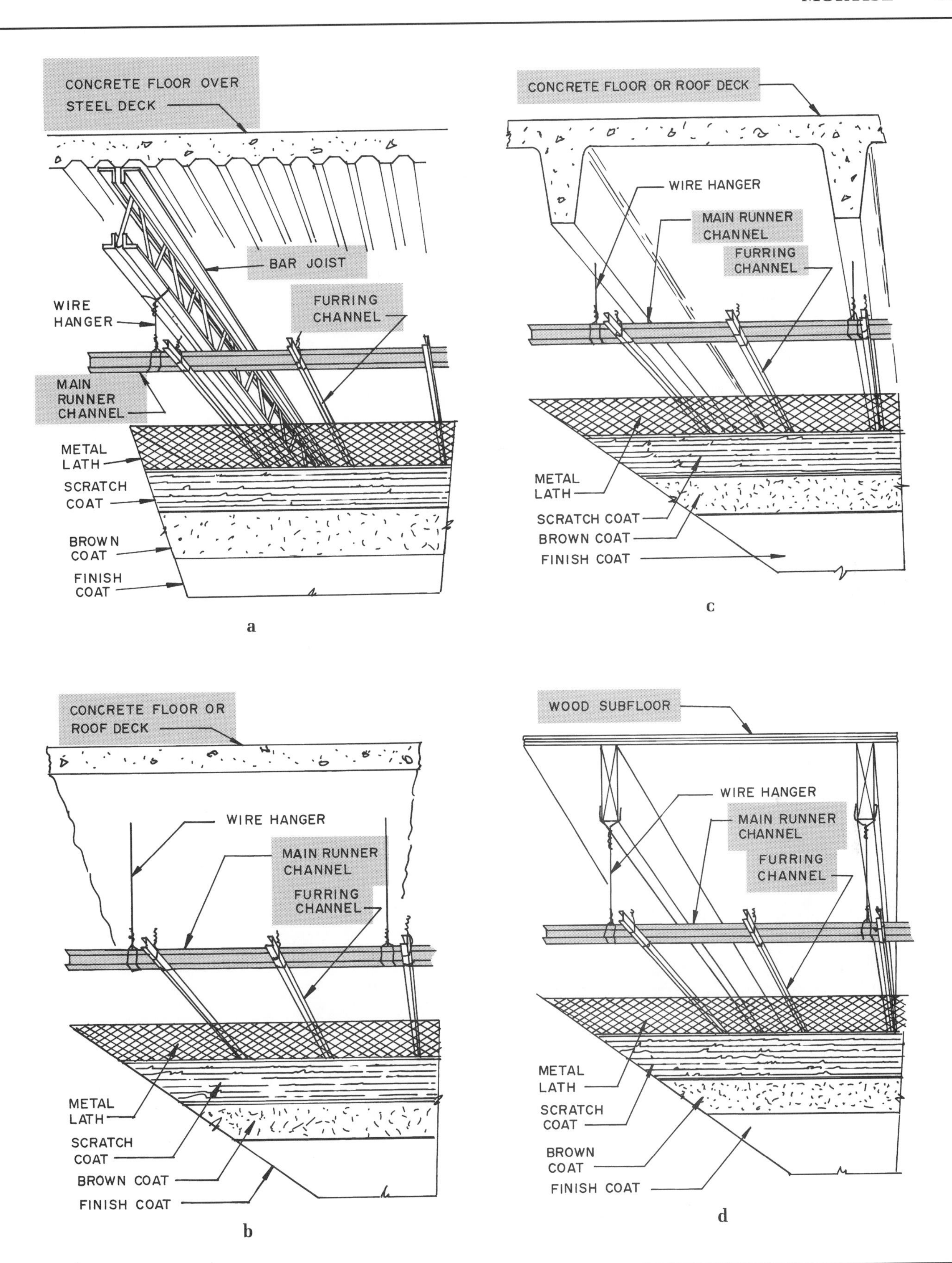

Figure M-5 Metal Lath.

Mortise and Tenon a type of wooden joint in which the projecting portion of one piece, the tenon, is fit tightly and glued into a recess cut in another, the mortise, which creates a structurally strong joint

Mud construction slang for mortar or gypsum wallboard joint compound

Mudsill pressure-treated board or redwood board fastened to the top of a foundation on which the rest of the building framing is erected; also called a *sill plate*

Mullion a slender bar, or divider, between two window units

Multiple Dwelling three or more dwellings that share a common roof and wall, such as in apartments or condominiums

Multiply Construction two or more layers or gypsum wallboard used to increase the fire rating of a structure

Muntin bars that are used to divide window openings; also called *sash bars*

Mushroom Cap the widened top of a column designed to better distribute the imposed load

Nail a slender metal rod with a pointed tip used to fasten materials together, typically wood and masonry; nails have a variety of different head shapes and sizes, depending upon their intended use; most nails are made of steel, though they are also available in copper, brass, and aluminum; nail types include box, duplex, finish, casing, common, drywall, roofing, and tacks

National Fire Protection Association (NFPA) a nonprofit, nongovernmental association of firefighters, engineers, and other individuals interested in fire safety; NFPA publishes the *Fire Protection Handbook* (1st edition 1896, 20th edition 2008), the premier reference source for fire-protection and fire-prevention information; other influential documents include the National Electric Code (NEC) and Life Safety Code (NFPA 101); standards related to building construction include the following publications:

NFPA 220 "Types of Building Construction"

NFPA 256 "Standard Methods of Fire Test of Roof Coverings"

NFPA 241 "Standard for Safeguarding Construction, Alteration, and Demolition Operations"

NFPA 251 "Standard Method for Fire Tests of Building Construction and Materials"

NFPA 703 "Standards for Fire Retardant Impregnated Wood and Fire Retardant Coating for Building Materials"

NFPA 1620 "Pre-Incident Planning"

NFPA 5000 "Building Construction and Safety Code"

The last document is used by some jurisdictions; in addition, the Fire Analysis and Research Division's "One-Stop Data Shop " (www.nfpa.org/research) provides statistics on a variety of fire-related research

NFPA 220 National Fire Protection Association Standard on Types of Building Construction (Types I, II, III, IV, V); outlines the required fire-resistive ratings of building elements for the various types of construction (**Figure F-5** Comparison of the construction classifications of the model building codes)

National Institute of Standards and Technology (NIST) part of the U.S. Department of Commerce; developed the fire-dynamics simulator (FDS) software modeling tool, among other things, to understand and visually represent the dynamics of gas and fluid flow, enabling fire-protection engineers to create smoke scenarios in different construction designs; the NIST also investigates line-of-duty deaths along with NIOSH

Neoprene a synthetic rubber that is oil-resistant and flexible

Neutral Plane the area of a beam below which compressive stress is nonexistent

Newel/Newel Post the upright post that supports the stair railing at the top and bottom of the staircase and at the landing, if there is one

Nogging filling spaces between structural members with bricks

Nominal Size the size designation of width and depth in standard sawn lumber and glued-laminate lumber grades

Nominal Strength the capacity of a structure or member to resist the effects of loads, as determined by computations using specified material strengths, and dimensions and formulas derived from accepted principles of structural mechanics, or by field or laboratory test, allowing for modeling effects and differences between laboratory and field conditions

Nonbearing Wall a wall that is not a bearing wall and only supports its own weight

Noncentered Ridge a roof ridge that is offset with respect to the centerline of the outer walls of a structure

Noncombustible a material that is fire resistant and incapable of being burned; any material that passes ASTM E136-09 is considered noncombustible (see also *Type II Construction*)

Non-combustible Construction also known as Type II Construction, it is characterized by non-combustible, fire-resistive structural components with limited use of combustible materials (see also *Type II Construction* **(Figure T-11)**)

Noncombustible Material a material that, in the form in which it is used and under the conditions anticipated, will not ignite, burn, support combustion, or release flammable vapors when subjected to fire or heat; materials that pass ASTM E136 are considered noncombustible materials

Non–Load-Bearing Wall a wall that does not carry any of the loads of the structure, such as a divider or partition; because they support only their own weight, non–load-bearing walls may be moved, modified, or removed without affecting the structural stability of a building (see also *Walls* **(Figure W-01)**)

Nonrigid flexible, as in an electrical cable than can be easily bent

Nonstructural Elements those columns, beams, arches, or trusses that do not carry a load

Nonstructural Wall all walls other than bearing walls or shear walls

Oakum long, untwisted, hemp rope fibers used to pack the joint and retain molted lead in bell and spigot pipe joints

Oblique not parallel or perpendicular to a plane or line; having no right angle

Obtuse Angle an angle larger than 90 degrees but less than 180 degrees

Occlude to close up or block off, such as to close up a duct

Occupancy the purpose for which a building or structure is used or intended to be used **(Figure O-1)**

Occupancy Classifications the use for which a building or structure is designed

Occupancy Groups various groupings of building use that include the following:

A – Assembly (auditoriums, theaters, stadiums)

B – Business (offices, retail shops, restaurants)

E – Educational (schools, day care facilities)

F – Factories (manufacturing plants, mills)

H – Hazardous uses (facilities handling flammable or explosive materials)

I – Institutional (hospitals, nursing homes, reformatories)

R – Residential (homes, apartment buildings, hotels)

S – Storage (warehousing facilities)

Occupancy Load the number of occupants a room or building may safely hold

Occupant Load the total number of persons for which the means of egress of a building or portion thereof is designed; typically determined by the use of the space or as the maximum probable population of the space under consideration at any one time (California Fire Code, 2007)

Typical Hazards Associated with Occupancies

Occupancy	Type of Construction	Hazards
Residential	Type V, most common	Fire loading, truss construction, owner alterations, rapid fire extension in void spaces
Commercial	Type III, most common	Fire loading, truss construction, rapid fire extension in void spaces, unknown occupancy change
Educational	Type II, most common	Unprotected structural steel, collapse, high fire load in some areas
Business	Types II and III, most common	Unknown change in occupancy, high fire load, difficult to ventilate
Industrial	Types I and II, most common	Hazardous materials, difficult to ventilate

Figure O-1 Typical Hazards Associated with Occupancies.

Occupancy Separations fire-resistive vertical or horizontal constructions required to prevent the spread of fire from one occupancy to another in a mixed-use building

Occupiable Space a room or enclosed space designed for human occupancy in which individuals congregate and which is equipped with means of egress, light, and ventilation

Octagon a polygon with eight sides and eight angles

Off-Center not centered; not located at an equal distance from edges of reference points

Offset Studs a framing method in which the studs are attached alternately on one side of the plate or the other so that drywall panels on each side of a wall are fastened to different studs

On Center (OC) the distance between items as measured from the centerline of one item to the centerline of the next; for example, for studs placed 16 inches on center, measure 16 inches from the centerline of one stud to the centerline of the next

One-Way Ribbed Structural Slab a structural slab with hollow filler blocks made of lightweight concrete, clay tile, or gypsum tile arranged in rows, or with metal pan fillers between concrete joists that result in a concrete slab with reduced dead load, as opposed to a solid slab that has equal load capacity; the depth of a one-way ribbed slab is greater than a solid slab, with 2 inches or more of solid concrete placed over the blocks to provide concealment space for piping and conduit and to add strength

One-Way Slab/Joist Slab a one-way slab is uniformly thick, reinforced in one direction, and cast integrally with parallel supporting beams; one-way slabs are suitable for light to moderate loads over relatively short spans (6 to 8 feet) and are supported on two sides by beams or load bearing walls; one-way joist slabs are cast integrally with a series of closely spaced joists, which are supported by a parallel set of beams; designed as a series of T-beams, joist slabs are more suitable for longer spans (15 to 36 feet) and heavier loads than one-way slabs. Tensile reinforcement occurs in the ribs

One-Way Structural Slab a slab with a uniform depth, and no filler material, that is reinforced in only one direction; this is an economical construction used for spans of up to 12 feet, where heavy concentrated loads will be carried

Openings apertures or holes in the building envelope that allow air to flow through the building and which are designed as "open" during design winds

Open Valley a type of roof installation in which the valley flashing is left exposed rather than being covered over by shingles

Open-Web Joist lightweight, shop-fabricated steel members for floors and roofs having a trussed web of steel bars or tubes, typically a steel joist that is built up using lengths of thin, lightweight structural steel shaped into a lattice pattern to make it less costly than solid steel joists and beams (steel trusses are used for heavy loads or long spans from 20 to 200 feet); cross bridging connects the top chords typically every 8 feet to prevent rotation or buckling and to resist lateral movement; also known as *open-web bar joist* **(Figure O-2)** (see also *Steel Joist Framing*)

Figure O-2 Open Web Joist.

Ordinance a rule, regulation, or law set by governmental authority

Ordinary Construction construction that uses masonry exterior walls and other structural elements wholly or partly of wood **(Figure O-3)** (see also *Type III Construction* **(Figure T-12)**)

Figure O-3 Ordinary Construction. *(Courtesy of Pat McAuliff.)*

Ordinary Combustibles materials in buildings that include paper products, furnishings, and other typical items that yield about 8000 BTUs per pound when burned

Oriented Strandboard (OSB) a nonveneered wood panel product commonly used for sheathing and as subflooring, made by bonding layers of long, thick wood strands under heat and pressure using a waterproof adhesive; the surface strands are aligned parallel to the long axis of the panel, making the panel stronger along its length; this engineered wood product has replaced plywood in many applications **(Figure O-4)**

Figure O-4 Oriented Strandboard (OSB).

Orifice an opening, mouth, or constriction in a passageway

⚠ **Out of Plumb** not truly vertical, such as a wall that leans slightly **(Figure O-5)**

Out of Square not square; a condition in which a true 90-degree angle does not exist but should; during construction, floors, walls, openings, doors, and forms for foundations are checked to ensure that they are square to save time throughout the construction process

Figure O-5 Out of Plumb.

Outrigger the rafter extension and structural support for the roof overhang, a design feature that provides shade, water runoff, and some degree of weather protection for the building

Overhang the projection of the second story of a building beyond the exterior wall of the first story and/or the length a rafter extends beyond a building's exterior wall

Overhead Door a large door that is raised overhead to open

Overlap to install a material with the bottom edge of one layer over the top edge of the next

Overlay a thin layer of material bonded to a panel or board for decorative or protective purposes

Paint a coating, made up of a combination of pigments and a binder, or a vehicle to carry the pigment; paint is designed to cover color and protect the surface to which it is applied, and most paints have either an oil or latex base and can be mixed in an unlimited variety of colors

Palladian Window a set of three windows, the top of the center one forming an arch, with a rectangular window on either side

Panelized Roof a roof composed of laminated beams, purlins, and two-by-four joists supporting a plywood deck covered with asphalt or tarpaper; primarily used in commercial construction for large buildings, such as supermarkets and convenience stores **(Figure P-1)**

Figure P-1 Panalized Roof.

Pane a section of window glass

Panel the space within the web of a truss, between any two panel points on a chord and a corresponding pair of joints, or a single joint on an opposite chord; also, any relatively thin, flat material used in construction, such as plywood panels for flooring, gypsum wallboard for walls, and acoustic panels for drop ceilings; a panel floor refers to the area of a floor between four supporting columns

Panel Point a joint between two or more members of a truss, panel points connect a principal web member and a chord and provide axial tension or compression loads to pass through those points

Panel Length the space on the chord of a truss between any two adjacent joints made by principal web members with the chord

Panel Wall a non–load-bearing exterior wall supported by the outer frame of a structure, which carries no vertical loads other than its own weight **(Figure P-2)** (see also *Curtain Wall*)

Figure P-2 Panel Wall. *(Courtesy of Precast/Prestressed Concrete Institute.)*

Panelized Construction construction that uses preassembled panels for walls, floors, and roofing

Panic Hardware a door-latching assembly incorporating an unlatching device, the activation portion of which extends across at least one half the width of the door leaf on which it is installed; installed in the direction of egress travel and designed to push open with minimal force

Parallam a trademark for a parallel strand lumber product used as beams, columns, headers, and lintels

Parallel-chord Wood Trusses a prefabricated, engineered structural element typically used for floor and ceiling assemblies; web members may be wood with metal gusset plates or steel tubes bolted into the upper and lower parallel wood chords; often used in residential construction, these trusses lessen the weight of wood beams while providing uniformity and greater load-carrying capacity and allowing for longer spans **(Figure P-3)**

Figure P-3 Parallel Chord Wood Trusses.

Parallel-strand Lumber a structural lumber product made by bonding long, narrow wood strands together under heat and pressure using a waterproof adhesive; may be used as beams, columns, headers, and lintels under the trademark Parallam

⚠ **Parapet** a low wall that rises above the roofline, it provides a decorative appearance as well as providing some protection from the elements; parapet walls do not withstand lateral loads very well, and any flex can cause them to fail or collapse **(Figure P-4)**

Figure P-4 Parapet.

Particleboard a nonveneered wood panel product made by bonding small wood particles under heat and pressure with resin; commonly used as a core material for decorative panels and cabinetwork and as underlayment for floors; also known as *chipboard*, particleboard is relatively strong, inexpensive, and is available in panels or sheets of varying thickness and sizes

Partition a nonstructural interior object that separates two areas or spaces

Partition Wall an interior wall that subdivides the space within a building of not more than one story in height; partitions may be structural (load bearing) or non–load-bearing; load-bearing partitions are an integral part of the building structure and carry part of the structural load; they cannot be removed or relocated unless suitable replacement load paths, such as structural beams or headers, are integrated into the structure to replace their function; load-bearing partitions may be wood or steel studs, reinforced masonry, or other structural components that serve as load paths; non–load-bearing partitions do not carry part of the structural load and may be moved, replaced, or modified without structural changes to the building

Party Wall a wall that is common between two separate living units, such as between apartments or condominium units; typically a bearing wall that stands on the property line separating and supporting two adjacent buildings and common to both

Pavilion a covered area with open sides; a canopy, large tent, or small ornamental building in a garden or park area used for shelter or entertaining

Peak Roof a sloping roof that points upwards; examples include gable, gambrel, mansard, shed, and hip roofs

Pedestal a vertical support or column

Pedestal Floor a floor system in which removable floor panels are held on pedestals in order to provide a crawl space or a space for running wires, cable, or ventilation or heating ducts under the floor

Pediment a gable pattern over a door or window

Pendant a hanging decoration or decorative ornament that points downward

Pendentive one of four triangular, concave braces joined together at the top to support a dome over a square structure; the braces form arches from one corner to another around the top of the structure

⚠ **Penetration** an opening made through one side of a fire-resistance–rated assembly ("membrane penetration") or an opening that passes through both sides of a vertical or horizontal fire-resistance–rated assembly **(Figure P-5a and b)**

Pentagon a five-sided plane geometric figure with equal sides and equal angles

Penthouse an enclosed, unoccupied structure above the roof of a building—other than a tank, tower, spire, dome, cupola, or bulkhead—that occupies not more than one third of the roof area and not rising more than twelve feet above a roof; often used to cover stairway openings at the roof level or to conceal HVAC or other building equipment

Perforated full of holes

Performance-based Option within some codes, assigns an objective to be met and establishes criteria for determining compliance, a relatively recent change in building codes that does not specify specific details on construction materials; rather, as long as safety from fire, structural failure, and other performance criteria are met, local jurisdictions may allow alternate methods of construction

Pergola a horizontal trellis mounted on columns designed to be an overhead covering for an open-air structure

P

Perimeter the outside boundary of a geometric shape or of a building or piece of property

Perimeter Ducting HVAC ducting that is routed around through the inside of a building's exterior walls; it may be located behind a false drop ceiling or disguised by some other method

Perimeter Foundation the portion of a building that is in direct contact with the soil and on which the building rests; unlike a slab foundation that covers the entire area under the building, a perimeter foundation extends only around the exterior

Perlite a lightweight, expanded mineral product used in insulation and as a lightweight aggregate

Permeablity a measure of the ability of water to flow through a material, such as concrete

Perpend a stone or masonry unit that passes completely through the thickness of a wall, such as a brick header extending from the interior surface to the exterior surface of a wall

a

b

Figure P-5 Penetration.

Perpendicular a surface or object at a right angle (90 degrees) to another surface or object

Pier a short, rectangular column with a large cross section often used to support beams, foundations, or decks; it may be wider at the bottom to increase its load-bearing area; also a supporting section of wall between two openings

Pier Footing a foundation footing for a pier or column

Pilaster a rectangular column embedded in and projecting slightly from one or both faces of a masonry wall; in addition to carrying vertical concentrated loads, such as girders or timbers, pilasters provide lateral support for masonry walls; pilasters may extend about one third its width from the wall

Pile or Piling a vertical structural foundation member that is driven into the ground and on which a foundation rests, pilings are used where surface soil conditions are not sufficiently stable to support the weight of a building; they are made of wood, concrete, steel, or a composite of concrete and wood, each offering different properties for various applications

Pillar a slender column or vertical structure of masonry or other material used to support a load or stand alone as a monument

Pinned a frame whose elements are connected by simple connectors, such as bolts or rivets

Pin Joint a structural connection that allows rotation but resists translation in any direction; also called *hinge joint* and *pinned connection*

Pitch a slope, grade, or incline, such as the angle of the slope of a roof, which is expressed as a ratio of the rise to span

Pitch Wall a wall with an upper plate that slopes to match the roof slope

Plain Concrete concrete that is unreinforced or contains less reinforcement than the minimum amount specified by code for reinforced concrete

Plancher Cut a horizontal trim cut on the tail of a rafter

Plancier the underside of an eave or cornice; often a horizontal plywood surface

Plank a rectangular cross-section board more than 6 inches wide and more than 1 inch thick; when referred to in floor and roof decking, planks are 2-inch wood

Plank-and-Beam Framing a floor or roof system that uses a supporting grid of wood posts or columns to form a skeletal frame structure; structural loads are concentrated on fewer and larger members than conventional framing; the structural system is usually left exposed (no concealed spaces) and is used to create large, unobstructed areas in residential, commercial, and industrial buildings; timber, steel, or concrete columns, timber or steel girders, or concrete or masonry bearing walls may support wooden beams; tongue-and-groove wood planks typically span 4 to 8 feet on top of the heavier beams; other planking/decking options include cementitious roof planks, 2-4-1 plywood, and/or prefabricated composite or stressed-skin panels; plank-and-beam framing may qualify as heavy timber construction, if the structure is supported by noncombustible, fire-resistive exterior walls and the members and decking meet the minimum size requirements specified in the building code

Plaster a mixture of lime, sand, and water used for coating walls and ceilings; it is mixed to a doughy consistency and cures to a hard surface

Plaster, Exposed Aggregate a mixture of cement, sand, and lime, commonly known as *stucco*, and used primarily as an exterior wall covering

Plaster, Gypsum calcined gypsum powder and/or lime with sand or other additives that can be worked as a paste with the addition of water; it sets up hard when dry and is used to plaster interior surfaces

Plaster, Portland Cement a plaster wall finish for interior or exterior use, consisting of Portland cement binder with aggregates

Plastic any of a number of synthetic compounds made from petroleum products that can be formed into desired shapes by pressure, heat, or extrusion and that harden or solidify under the proper conditions; plastics contain approximately 16,000 BTUs per pound

Plate the top or bottom horizontal structural member of a frame wall or partition; the sole plate is the bottom horizontal member of a framed wall upon which a row of studs is erected, and the top plate connects beams like joists and rafters

Plate Cut a bird's mouth or angle cut out of a rafter to allow the rafter to sit back into the top plate of a wall, rather than balance on top of it (see also *Bird's Mouth* **(Figure B-7)**)

Plate Girder a structural member built up of steel plates and angle iron fastened together to form a strong, yet relatively lightweight section

Platform a raised floor area within a building commonly used as a stage area for speakers or for the presentation of music, plays, or other entertainment wherein there are no overhead hanging curtains, drops, scenery, or stage effects other than lighting and sound

Platform Framing a light wooden frame having studs only one story high, regardless of the stories built; each story rests on the top plates of the story below or on the sill plates of the foundation wall to provide fire stopping at each level; typical stud spacing is 16 or 24 inches; platform framing is also referred to as *Western framing.* The other major type of wood framing is *balloon framing,* in which studs run from the bottom of the first floor to the top of the second floor **(Figure P-6)**

Platform Stairway a stairway with a change of direction or landing between floors

Plenum a compartment or chamber to which one or more air ducts are connected that forms part of the

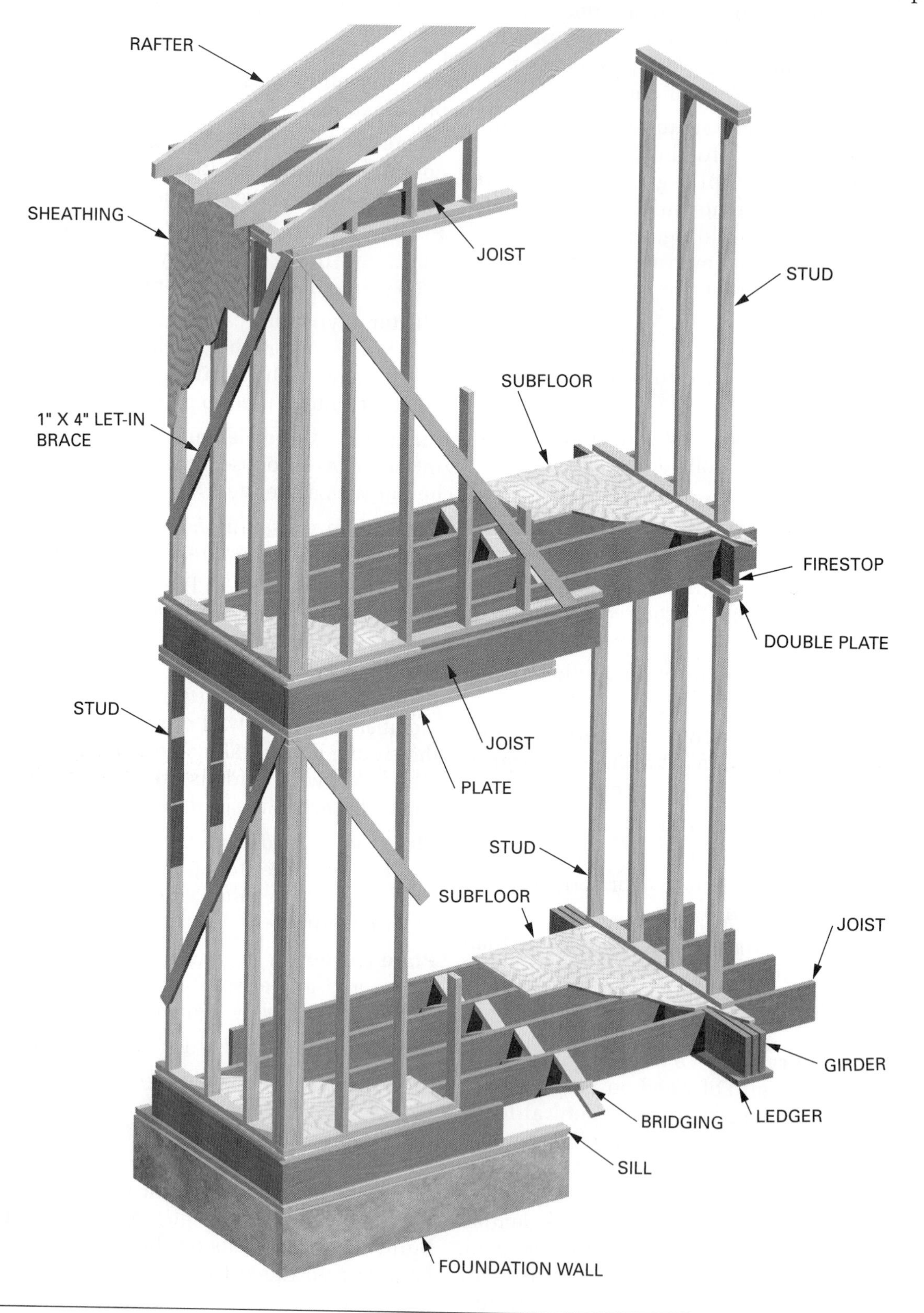

Figure P-6 Platform Framing.

air-distribution system; may also be the void space in a floor or ceiling assembly or above a suspended ceiling

Pliable flexible, bendable, or moldable, as in products made to be pressed or molded into position for use, such as duct-seal putty, weather-stripping, and caulk

Plumb true vertical; to be or cause to be vertically aligned

Plumb Cut a vertical cut that is exactly straight up and down; plumb cuts are made at the top end of a rafter, where it meets the ridge

Plywood a wood panel product made from several veneer layers bonded together under heat and pressure, usually with the grain at right angles to each other; two main types of plywood are *exterior grade plywood* and *interior grade plywood*, which are graded A through D according to its face veneers; engineered grades of plywood are used for wall and roof sheathing, subflooring, underlayment, and interior finish; a panel-grade stamp identifies the intended use or veneer grade (A, B, C, D, or N) of a wood panel product; the term *2-4-1 plywood* refers to 1-1/8-inch plywood, which provides shear strength. When aflame, plywood delaminates readily, increasing the surface area and the rate of heat release; oriented strand board (OSB) is used to replace plywood in many applications

Plywood, Exterior plywood made with waterproof glue, used for exterior siding, roof sheathing, and other applications where waterproof glue is essential to retain the integrity of the material

Plywood Grades ratings that indicate the quality of the plywood, ranging from Grade 1, which has matched grains and can be used for furniture or panels, to Grade 4, which has any type and number of defects but is sound enough to be used in construction

Plywood Interior plywood made with water-resistant glue that can be exposed to occasional or moderate amounts of moisture but is not suited to exterior applications

Plywood Veneer plywood that is made of several layers of veneer glued together and in which the outer layers are of an expensive or higher quality wood, such as walnut or oak

Pokethrough an opening made in a wall, floor, or ceiling to accommodate utility services **(Figure P-7)**

Porch a part of the structure that projects out from the exterior walls to provide a covered entrance to a house or a covered recreation or seating area outside of the main structure

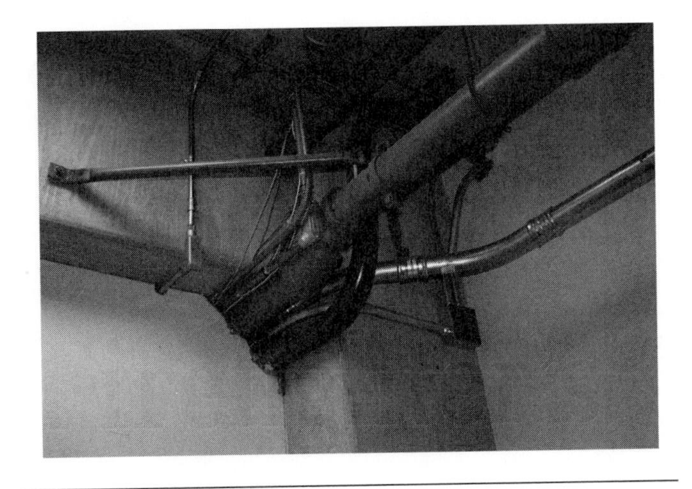

Figure P-7 Poke Through.

Portal an entryway, door, or opening

Portico a covered porch or walkway supported by columns, usually extending out from a building

Portland Cement a material used as a base for mortar and concrete; it is a ground and calcined mixture of shells, limestone, cement, rock, clay, marble, shale, sand, and iron ore; when mixed with water, it undergoes a chemical reaction resulting in a hard, strong, homogeneous structural material

Post a stiff, slender vertical support member; in timber framing, refers to a wooden column

Post-and-Beam Framing a construction technique that utilizes a framework of vertical posts and horizontal beams to carry both floor and roof loads; the beams supporting the floor and roof systems transmit their loads to posts or columns that, in turn, carry the loads down to the foundation system; together with plank-and-beam floor and roof systems, the post-and-beam wall system forms a three-dimensional structural grid or framework within which nonbearing wall panels, doors, and windows are integrated; wood post–beam connections include shear plate or split-ring connectors, exposed column caps, exposed T-straps, continuous posts, continuous beams, space posts, concealed connections, and interlocking connections; often post-and-beam framing is left exposed and creates large, unobstructed open areas in a building; post-and-beam framing may qualify as heavy timber construction if the plank-and-beam floor and roof structures are supported by noncombustible, fire-resistive exterior walls, and the wood members and decking meet the minimum size requirements specified in the building code **(Figure P-8)**

Post and Frame a type of building construction identified by a frame or skeleton of timber fitted together; the principle members are of large

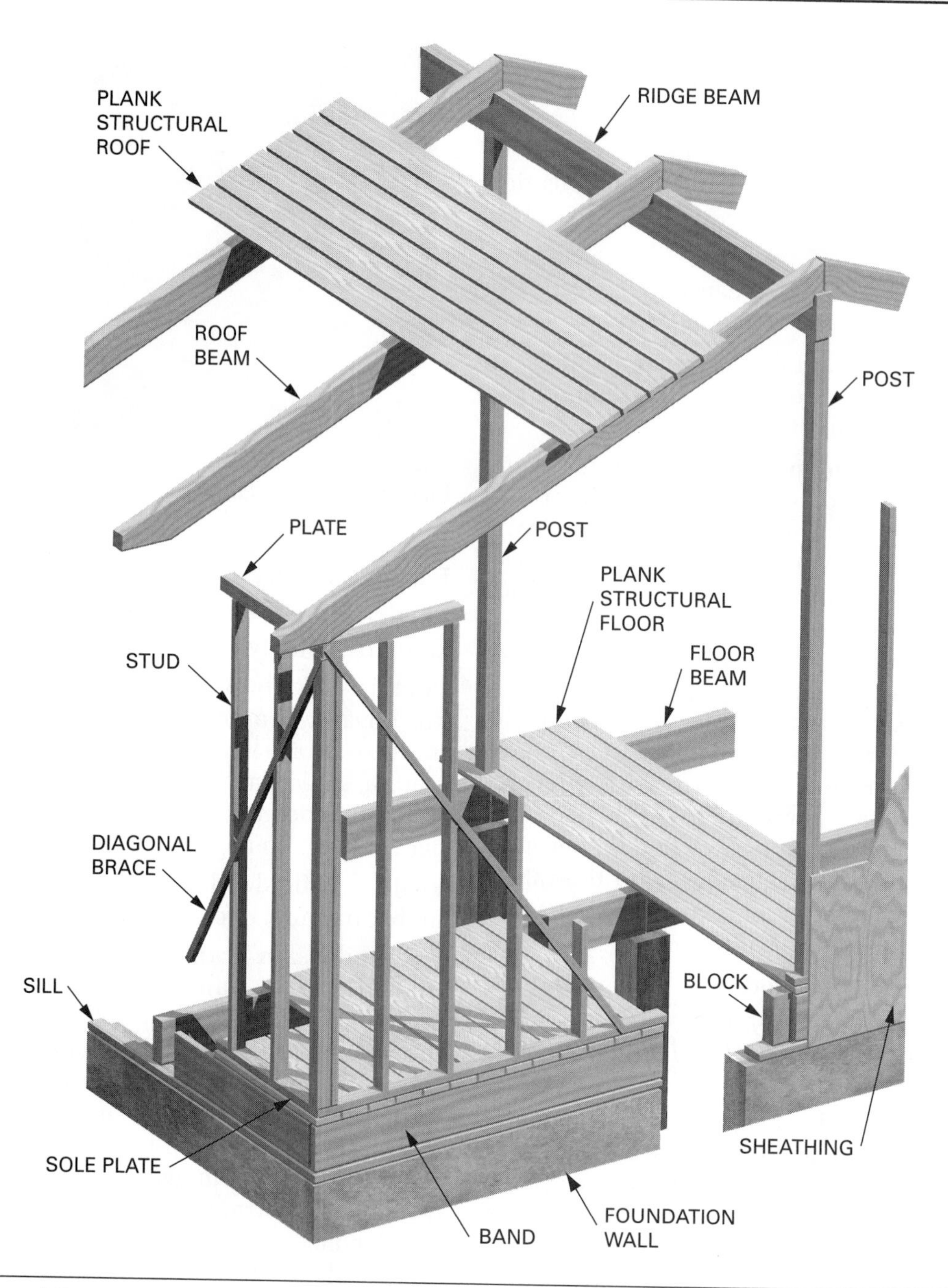

Figure P-8 Post-and-Beam Framing.

P

dimension, and smaller members between joints are constructed by mortise and tenon (socket and tongue) and pinned with trunnels (wood pegs)

Post and Girt see *Post and Beam*

Post Tensioning the prestressing of a concrete member by tensioning the reinforcing tendons after the concrete has set and then anchoring it at the ends of the structure to maintain that engineered strength; steel cables are drawn through holes in a concrete slab or other concrete structure, and tension is applied to the cables after the concrete has reached a specified strength; the tension load is applied to the concrete by means of plates attached to the cables. Concrete is strongest in compression, and this increases the strength of the overall concrete structure by putting it under compression stress.

Post Tensioned Concrete Structures generally site-fabricated, large-area buildings that use bonded (grouted) and unbonded tendons to create engineered strength by tensioning; typically seen in high-rise construction and parking garages, posttensioning creates greater strength without adding weight to the structure; the use of water-resistant grease contained in a protective sheath, and anchorage covers at the end fittings of tendons, helps provide for corrosion resistance **(Figure P-9)**

Pounds Per Square Inch (PSI) a measure of pressure, as a load applied by one commodity on another

Poured-in-Place mortar or concrete that is poured and cured at its permanent location and position; perimeter and slab building foundations are examples of poured-in-place concrete poured in forms at the building site (also known as *Cast in Place* **(Figure C-02)**)

Precast Concrete concrete items (cinder and concrete blocks, beams, columns, panels) cast in their final shape in forms and then brought to the construction site for installation; the strength, density, and quality of the concrete are almost always superior to that achieved with site-cast concrete; precast architectural concrete is used as part of the structural framework and/or siding of a building; precast wall units include curtain or load-bearing walls, shear walls, wall support units, and exterior forms

Precast Concrete Floor and Roof Systems precast concrete slabs, beams, and structural tees are one-way spanning units that may be supported by site-cast concrete, precast concrete, or masonry bearing walls, or by steel, site-cast concrete, or precast concrete frames; the precast units are manufactured with normal-density to structural lightweight concrete and are prestressed for greater structural efficiency, which results in less depth, reduced weight, and longer spans; precast concrete units include solid flat slabs (4- to 8-inch depth, typical 4-foot widths, *and* 12 to 24-foot spans); hollow core slabs (6- to 12-inch depth, 1- to 8-foot typical widths, amd 12- to 40-foot spans); single tees (20- to 48-inch depth, 8 and 10 foot widths, 30- to 120-foot spans); double tees (12- to 32-inch depth, 8- and 10-foot widths, and 30- to 100-foot spans); beams (rectangular, L-shaped, and inverted T-beams for spans 15 to 75 feet); and AASHTO Girders, typically used for bridge structures but sometimes for building construction (36- to 60-foot spans)

Precast Contrete Frame Buildings one of the four construction classifications used by FEMA USAR teams to describe buildings for postfailure operations (others are *heavy wall, heavy floor,* and *light frame*); these buildings use modular, precast, and reinforced concrete structural elements for floors, walls, and columns and are assembled on-site; the concrete load-bearing frame uses tees with precast girders and columns that are typically one to ten stories high; building uses include commercial, mercantile, office, multiuse facilities, stadiums, and parking structures **(Figure P-10)**

Precast Concrete Wall Panels solid, composite, or ribbed panels that are cast in a plant off-site and transported to the construction site; they are conventionally reinforced or prestressed for greater structural efficiency, reduced panel thickness, and longer spans; they may also serve as bearing walls capable of supporting site-cast concrete or steel floor and roof systems; together with precast concrete columns, beams, and slabs, the wall panels form an entirely precast structural system that is inherently modular and fire resistive; the wall panels, in turn, must be stabilized by columns or cross walls, as they transfer the lateral forces to the ground foundation; all forces are transferred by a combination of grouted joints, shear keys, mechanical connectors, steel reinforcement, and reinforced concrete toppings; continuity between columns, beams, slabs, and walls is easily attainable in concrete construction

Precast Prestressed Concrete Panels wall panels that are erected in large sections in some steel-framed buildings

Preengineered Steel Building a steel-framed building covered with sheet steel; the building components are manufactured in a factory and shipped in pieces to the site, where they are erected and assembled on a foundation, forming a complete structure

Prefabricated modules, parts, or units made or constructed at a location, often at a factory, apart from the final place of installation; the prefabricated modules or units make the installation or assembly of the finished product easier, faster, and more uniform

Prefabricated Joists and Trusses commonly used in place of dimension lumber to frame floors, because they are generally lighter and more dimensionally stable than sawn lumber, and manufactured in greater depths and lengths to span longer distances; types include I-joists, wood chords, webs with gusset-plate connectors, and wood chords with steel webs or tubing webs **(Figure P-11)**

Prefabricated Metal Buildings normally one story, relatively lightweight structures in which framing is generally steel throughout, usually with metal decking and siding over light-gage steel purlins and girts; roofs normally have horizontal rod bracing with bolts

Prefire Anaylsis also referred to as *prefire plans, preplans, preincident planning,* or *target hazards.* Building information includes age, size (number of

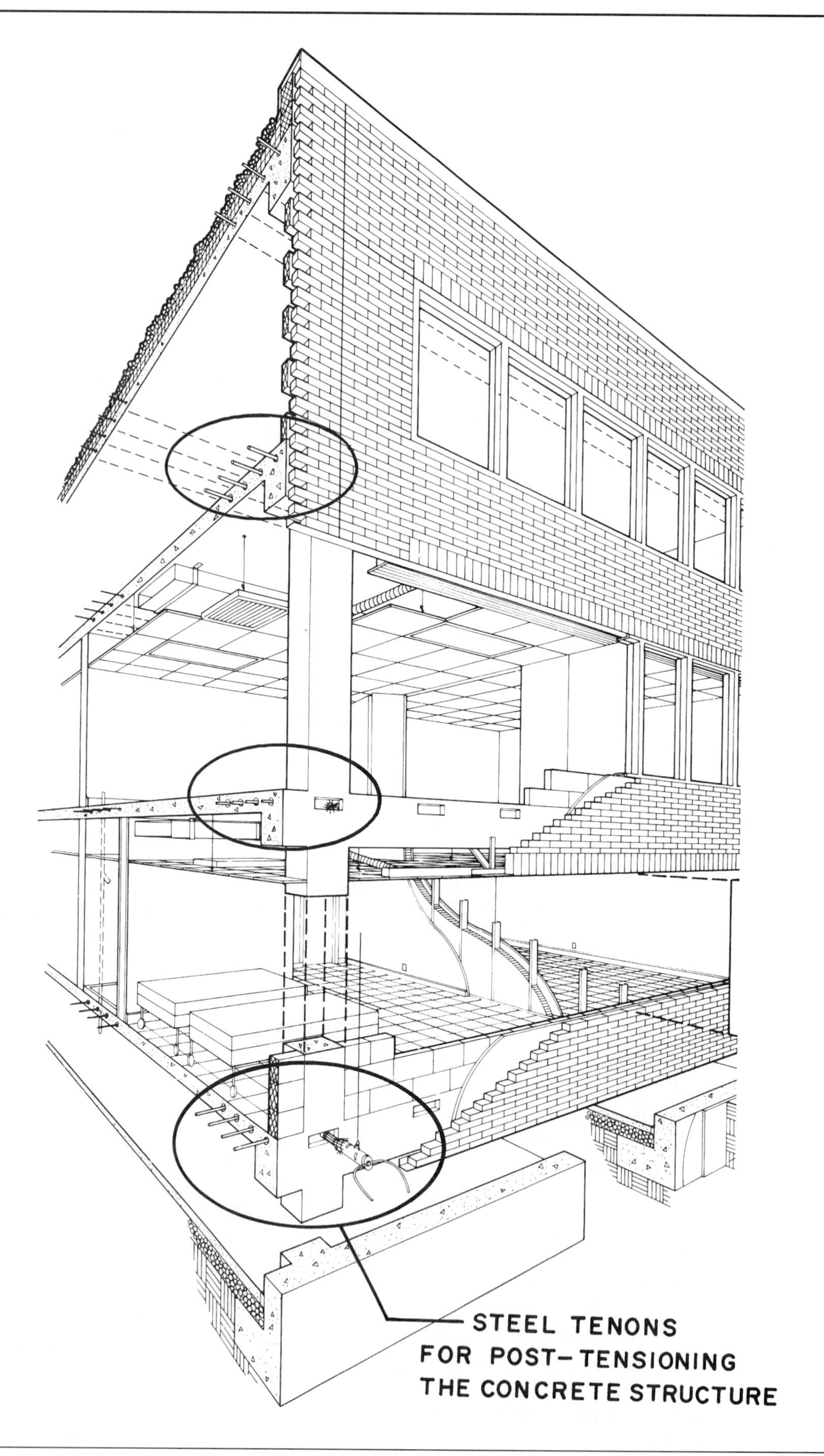

Figure P-9 Post Tensioned Concrete Structures.

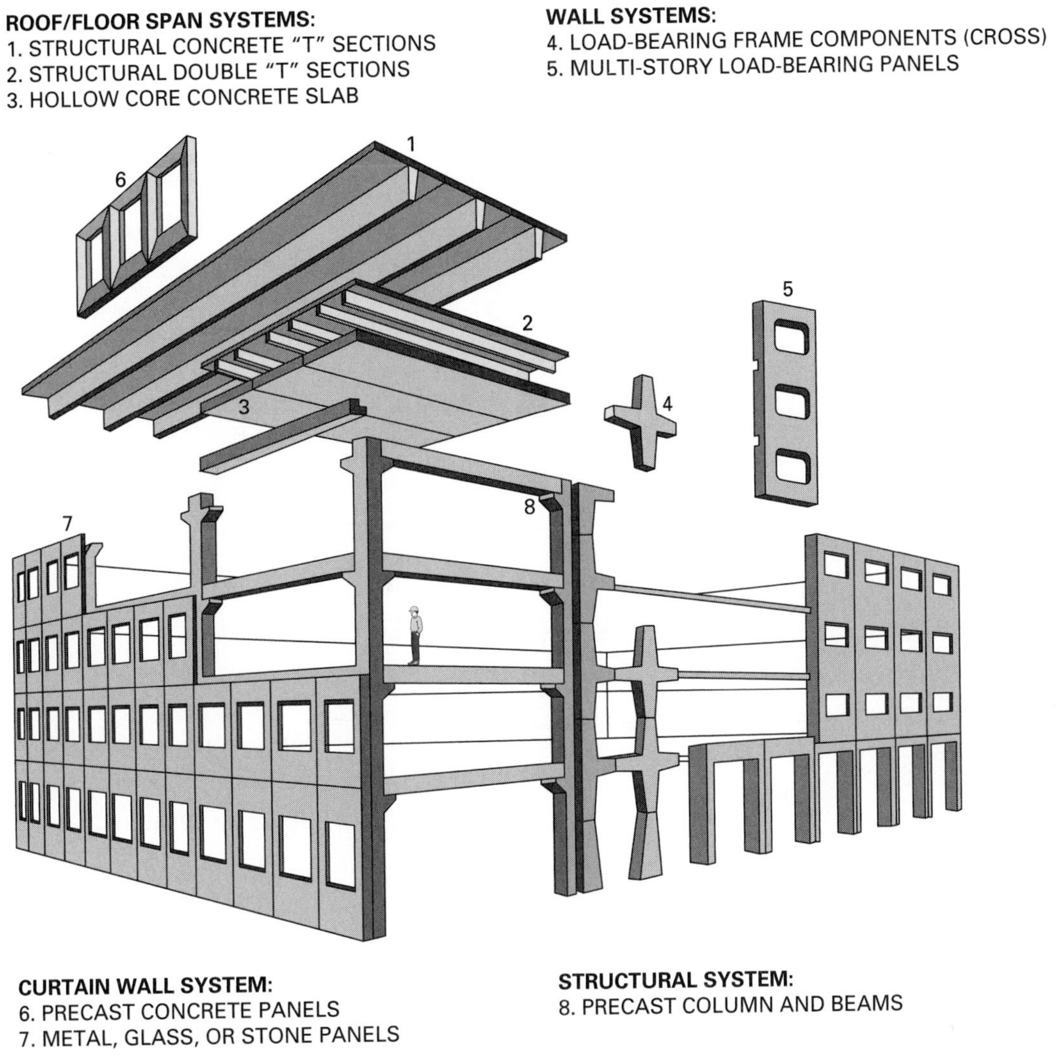

Figure P-10 Precast Concrete Frame Buildings. (*Courtesy of FEMA/University of California at Berkeley.*)

stories and square foot per floor area), occupancy and occupancy load, superstructure (frame or bearing wall), hazardous features, fire-protection systems, and the building's condition. NFPA 1620, "Preincident Planning," provides general guidelines

Prehung Door a door prefabricated with frame, hinges, and trim members intended to be installed as a unit

Preplanning prior determination of action to be taken in case of fire in a specific structure

Preservative-treated Wood wood impregnated under pressure with compounds that reduce their susceptibility to deterioration caused by fungi, insects, or marine borers

Prestressed Concrete a method of concrete reinforcement by either pretensioning or posttensioning high-strength steel tendons, which may be in the form of wire cables, bundled strands, or bars; the tensile stresses in the tendons are transferred to the concrete, placing the entire cross section in compression; the prestressed member is able to deflect less, carry a greater load, or span a greater distance than a conventionally reinforced member of the same size, proportion, and weight, which allows for lightweight concrete structures of extremely long spans; alterations, maintenance, and demolition activities must be considered carefully due to the stresses locked into the concrete system; there are two types of prestressing techniques: Pretensioning is accomplished in a precasting plant, while posttensioning is usually performed at the building site, especially when the structural units are too large to transport from factory to site. Hydraulic jacks, anchorage hardware, and the sequencing of the prestressing operations are important components of prestressed concrete **(Figure P-12)**

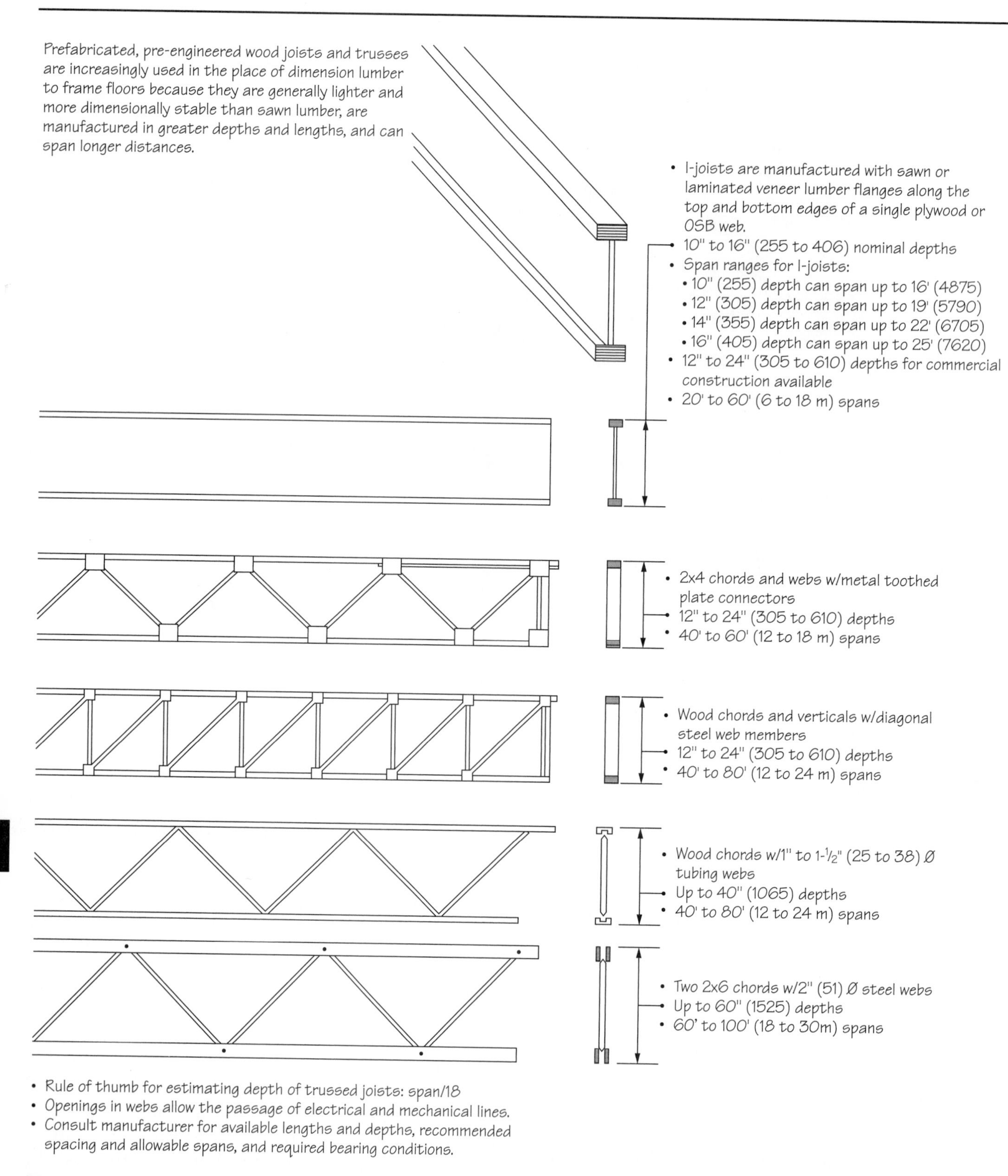

Figure P-11 Prefabricated Joists and Trusses. *(Courtesy of* Building Construction Illustrated, *Third Edition, by* Francis D.K. Ching, *© 2008, Reprinted with permission of John Wiley & Sons, Inc.)*

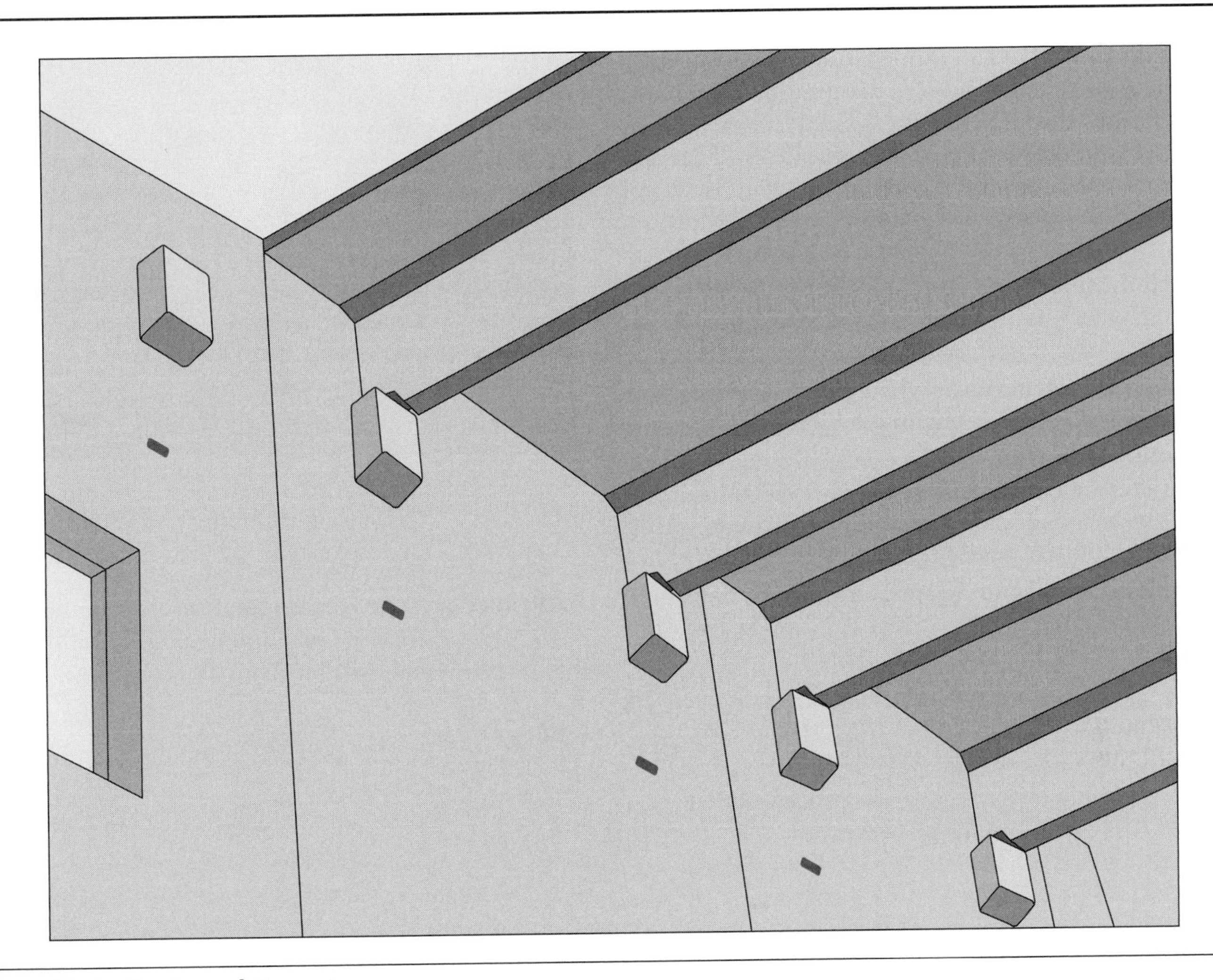

Figure P-12 Prestressed Concrete.

Prestressing the process of inducing compressive forces in concrete before the load is applied; steel in the concrete is tightened or preloaded, which creates compressive stresses in the concrete that counteract the tensile stresses that result when loads are applied; two methods are *pretensioning* and *posttensioning*; in pretensioning, steel cables or wires are stretched in the forms to put tension on them before the concrete is placed; the tension is kept on them until the concrete has set up; after the concrete hardens, the steel remains in tension, adding to the strength of the concrete; posttensioning is done on the job site, when steel tendons are stretched and anchored at the ends after the concrete has set

Pressure-treated Wood treated wood that resists insects and moisture; basement walls, mud sills, ledger boards, and decking commonly use pressure-treated wood

Pressurized Stairwell Enclousure a type of smoke-control system in which stairway enclosures are mechanically pressurized to minimize their smoke contamination during a fire incident

Pretensioned Concrete precast concrete subjected to internal compressive stress before being placed in position; pretensioned concrete is the result of tension being applied to reinforcing steel elements in a factory, prior to pouring the concrete; after the concrete is poured, it bonds to the reinforcing and maintains greater load-carrying capabilities

Pretensioning a type of prestressing of concrete accomplished by stretching the reinforcing tendons before the concrete is cast; steel tendons are first stretched between two abutments until a predetermined tensile force is developed; concrete is then cast in formwork around the stretched tendons and fully cured; when the tendons are cut or released, the tensile stresses in the tendons are transferred to the concrete through bond stresses, adding to the strength of the concrete; camber is produced, which is equalized by the deflection of the member under loading

PSI pounds per square inch

Projections on exterior walls, cornices, eave overhangs, exterior balconies, and similar projections extending beyond the floor area

Proscenium Wall the wall that separates the stage from the auditorium or assembly seating area; for stages or platforms where the height is greater than 50 feet, all portions of the stage must be completely separated from the seating area by a proscenium wall, with not less than a 2-hour fire-resistance rating extending continuously from the foundation to the roof

Protected Membrane Roof Assembly (PMRA) a design in which ballast, such as gravel, and insulation are applied over a roof membrane

Protected Construction when 1-hour fire-resistive construction is used throughout a building

Protected Opening an opening in a wall, floor, or roof-ceiling construction that is fitted with a fire assembly having the required fire-resistance rating for its location and use

Protective Membrane a surface material that forms the required outer layer or layers of a fire-resistive assembly containing concealed spaces

Purlin a horizontal member that acts as a beam to support common rafters or ceiling joists, set at right angles to rafters/trusses/joists **(Figure P-13)**

Figure P-13 Purlin. *(Courtesy of Kathleen Siegel.)*

Pyrolysis/Pyrolytic Decomposition decomposition of wood due to heat; an example would be a flue vent too close to wood members causing charring

Queen Post one of two secondary vertical tie posts in this specific type of truss

Quicklime calcium oxide, a caustic lime that is mixed with water to make a fine, puttylike substance used in finish coats, such as plaster (see also *Lime*)

Rack Storage a combination of vertical, horizontal, and diagonal members that support stored materials; may be fixed or portable **(Figure R-1)**

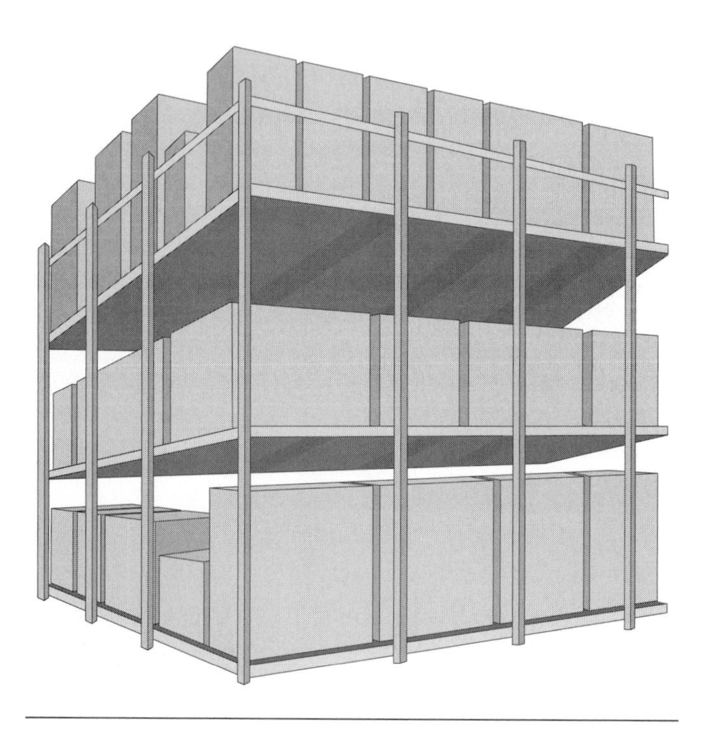

Figure R-1 Rack Storage.

Rafter an inclined beam that supports a roof; the main structural support member for the sloped roof of a building, extending from the peak of the roof to the eave line; rafters are designed to support the roof sheathing and covering as well as roof loads, and several different types of rafters are available to accommodate different roof styles; the common rafter extends from the peak of the roof to the eave live and is perpendicular to both the ridge and the wall plate to which it is attached; a roof with more than two slopes, such as a hip roof, will generally require hip and valley rafters; a hip rafter supports the intersection where two roof slopes slant up to meet at an external angle; the hip extends diagonally from the ridge to one of the outside corners of the wall; a valley rafter supports the inside intersection of two roof slopes and extends from the ridge to an inside corner of the exterior wall; jack rafters are short rafters that span the distance from a hip to the wall top plate or from a valley to a ridge; these rafters are often referred to as hip jacks or valley jacks to identify their location; cripple-jack rafters run between two other rafters, such as between a valley and a hip rafter, and do not bear on the ridge or wall plate; like other jack rafters, their names—*valley* cripple or *hip* cripple—designate their location **(Figure R-2)**

Rake the overhang of a roof on the gable end of a building

Raker a diagonal column, usually of wood, used as bracing

Rake Cornice the junction of the eaves on a gable roof end and the walls of the building

Rake Wall the wall of a building that has a sloping top plate, such as a structure with a shed roof

Rate of Heat Release (RHR) indicates how fast the potential heat in a fuel is released

Rating fire-protection rating is the designation indicating the duration of fire test exposure to which a door or window assembly was exposed and met all the acceptance criteria as determined with NFPS 252 or NFPS 257

Reaction the result of force exerted by a beam on a support

Rebar iron rod of various sizes used in concrete to give greater strength **(Figure R-3)** (see also *Reinforcing Bar*)

Rebar Ties wires used to connect two or more lengths of rebar, so they form a continuous length, or to tie lengths of rebar together where they cross

Recess a cavity or indentation

Redundancy duplication or repetition of elements in case of failure; examples include water-resisting system designs and load-carrying capabilities of structural members

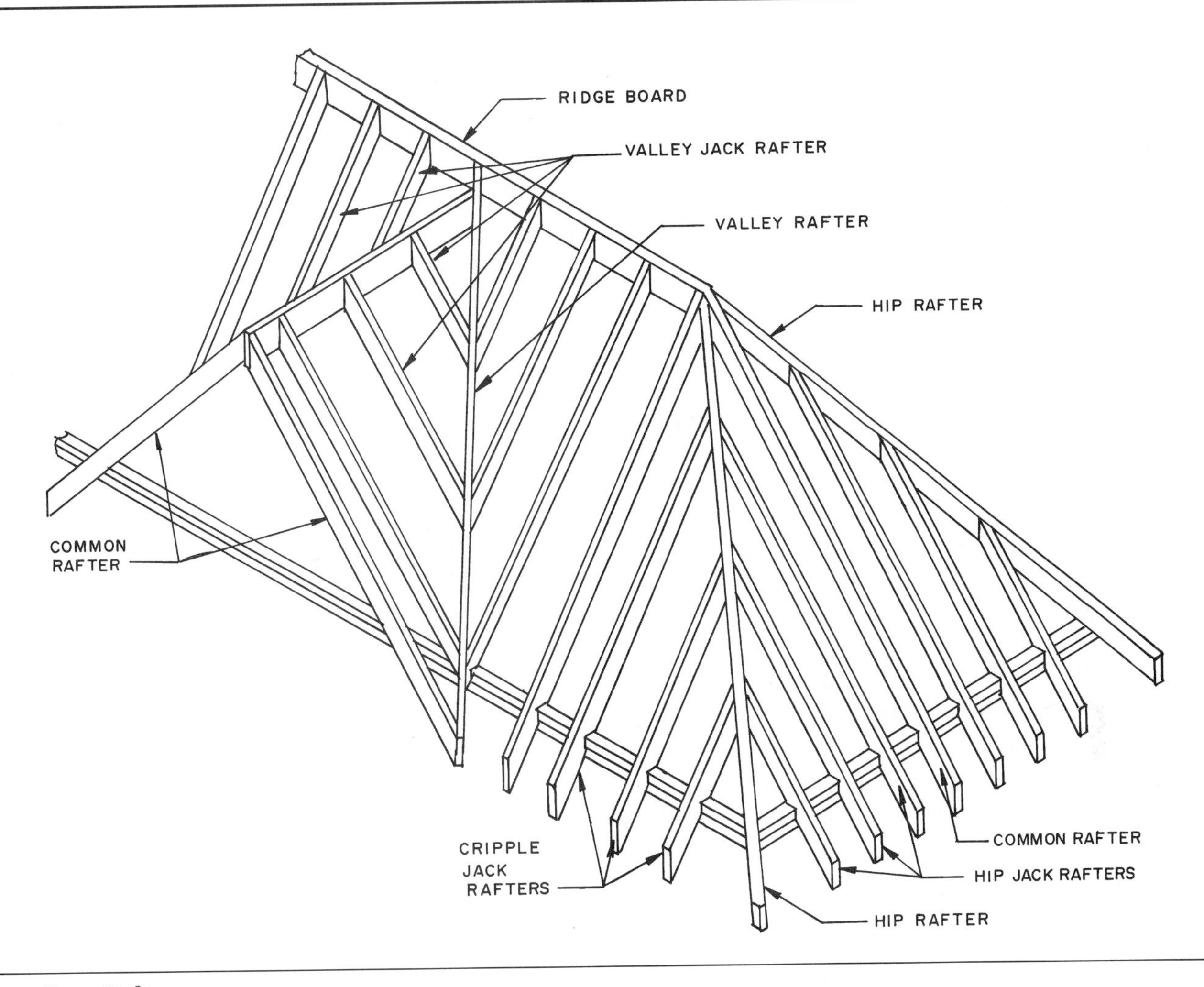

Figure R-2 Rafters.

Figure R-3 Rebar.

Rehabilitation in regards to buildings, rehabilitation refers to the removal and replacement of materials that have degraded

Reinforced Concrete concrete into which steel rods have been inserted to create a composite material with greater tensile strength

Reinforced Concrete concrete to which reinforcement has been added in the form of steel rods, bars, or mesh to increase its strength and resistance to cracking; a composite material in which steel rods or cables provide tensile strength, and concrete provides compressive strength; structural elements such as columns, beams, and wall and floor panels may be precast or cast in place; desirable characteristics of reinforced concrete include high compressive strength, good resistance to fire and water, rigidity, low maintenance requirements, economical material for below-grade structures, and the use of locally available inexpensive labor and materials; undesirable characteristics of reinforced concrete include composite material complexities; low tension and low shear strength without reinforcement; form, falsework, and shoring are required; low strength per unit weight; low strength per unit volume; wide variability; dimensional stability problems; and lack of ductility

Reinforced Concrete Beam a concrete beam designed to act together with longitudinal and web reinforcement in resisting applied forces

R

Reinforced Concrete Floor Systems depending upon the load conditions, span, and depth of the floor system, slab types include one-way slabs, two-way slabs, one-way ribbed slabs, two-way waffle slabs, two-way flat slabs, and two-way flat plates

Reinforced Construction a type of masonry wall construction common after 1940 that includes floor and roof diaphragms made of steel decking, with or without concrete fill; cast-in-place concrete, glulam and plywood; and precast planks or prestressed elements with concrete topping

Reinforced-faced Masonry walls of two widths of masonry units, each of different materials, such as concrete masonry units with glazed masonry unit facing bonded to it

Reinforced Masonry a structure of hollow concrete masonry units (CMU) or drilled masonry units reinforced with steel and mortar to resist forces; also called *reinforced filled-cell masonry,* unreinforced masonry (URM) lacks steel reinforcement, which makes it susceptible to failure during fires and earthquakes

Reinforcement structural strengtheners, such as steel reinforcing bars, used to add stiffness and/or strength to materials

Reinforcing Bar steel bar designed to be placed in concrete for reinforcement; concrete has compression strength; reinforcing bar adds resistance to breakage when other types of forces are applied; the bars have a patterned or textured exterior, which allows them to bond or interlock with the concrete mix; they come in a variety of sizes, starting at 1/4-inch diameter and increasing in size in eighth-inch increment (see also *Rebar* **(Figure R-03)**)

Renovation to restore to an earlier condition, as by repairing or remodeling

Repeated Loads loads that are applied intermittently

Reshores temporary shoring placed under concrete to help support it after the original forms have been stripped

Residential pertaining to residences or homes, as in residential construction

Resilience the ability of a material to be deformed under a load and return to its former size and shape when the load is removed

Resilient Tile floor tile made of vinyl, rubber, or other resilient material available in many sizes, colors, and patterns

Restrained Beam End the welded, nailed, bolted, or cemented end of a floor or roof beam

Retaining Wall a wall that is subjected to lateral loads other than wind loads, such as a wall used to hold back earth or water

Retrofit to furnish with new parts or equipment to bring a structure up to current standards

Return a surface that turns back on itself, or away from the primary or main surface, such as some types of decorative moldings

Reveal the exposed side of a door or window frame, between the face of the frame and the surface of the surrounding wall

Reinforced Concrete Columns usually cast with concrete beams to form a monolithic structure. They may also support timber or steel beams

Reinforced Masonry Walls modular building blocks that utilize steel reinforcing bars embedded in grout-filled joints and cavities to aid the masonry in resisting stresses

Reinforcing Bars also called *rebar,* steel sections of bars or rods hot-rolled with ribs or other deformations for better mechanical bonding to concrete; the bar number refers to its diameter in eighths of an inch: for example, a # 5 is 5/8 inches in diameter (see also *Rebar* **(Figure R-03)**)

Reinforcing Mesh a grid of thin steel rods that provide added strength for flat surfaces, such as slabs

Renovation the refinishing, replacement, bracing, strengthening, or upgrading of existing materials, elements, equipment, or fixtures without reconfiguration of spaces **(Figure R-4)**

Figure R-4 Rennovation.

Repair the patching, restoration, or painting of materials, elements, equipment, or fixtures for the purpose of maintaining them in good or sound condition

Reshoring in concrete construction, after falsework is removed during curing, shores are placed to help carry the load of the not-yet full-strength concrete

Resistance Factor a factor that accounts for deviations of the actual strength from the nominal strength in the manner and consequences of failures, also refered to as *strength-reduction factor*

Restraint the restriction of one member by another

Retaining Wall a wall that is not laterally supported at the top and is designed to resist lateral soil load

Return the continuation of a molding, projection, or other part at an angle, usually 90 degrees to the main part

Ribbed Slab a slab that is cast integrally with a series of closely spaced concrete joists supported by a parallel set of beams; used in one-way joist slab construction to span longer distances and carry heavier loads

Ribbon a piece of 1 × 4 lumber that is installed horizontally into recesses cut or "let" into the studs in balloon framing as a ribbon on which the second floor joists rest; also called a *ledger* or *girt* (see also *Ledger* or *Girt*)

Ridge the roof peak, the highest point of a sloped roof; may be viewed as a horizontal line of intersection at the top between two sloping planes

Ridge Beam a structural horizontal member supporting the upper ends of rafters at the ridge of a roof **(Figure R-5)**

Ridge Board the upper horizontal support member of a sloping roof against which the top ends of rafters are aligned and fastened; also called a *ridgepole*

Ridge Cap a roof covering along the ridge of a roof, often of continuous metal

Ridge Course the layer of shingles or other roofing material along the ridge of a roof

Ridge Cut the cut at the ridge end of a rafter

Ridgepole see *Ridge Board*

Ridge Vent A covered opening along the ridge board that provides ventilation for pitched roofs **(Figure R-6)**

Figure R-5 Ridge Beam.

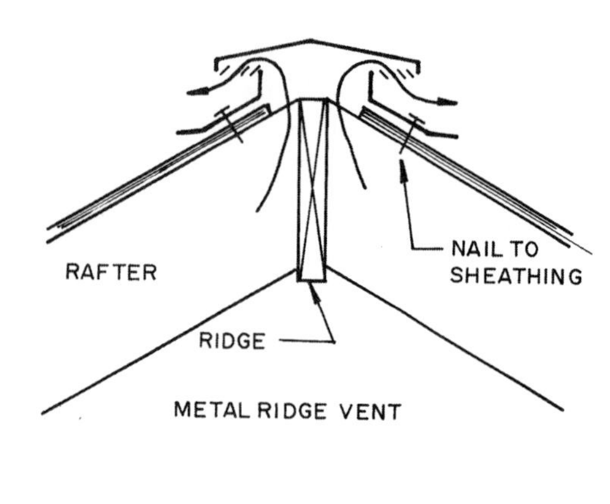

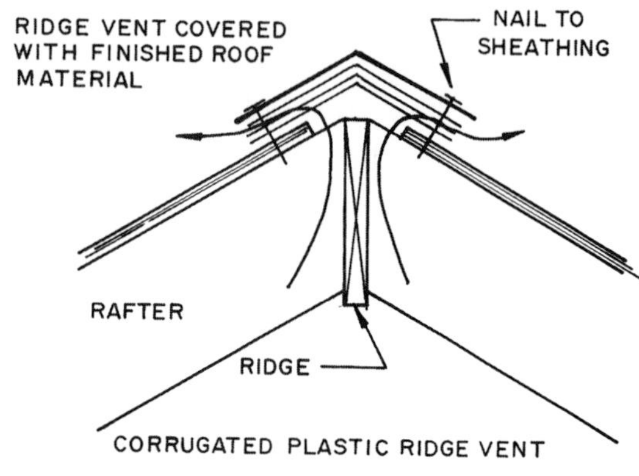

Figure R-6 Ridge Vent.

Right Angle an angle of 90 degrees

Rigid not flexible

Rigid Base Diaphram a reinforced concrete slab placed on the ground

Rigid Foam Insulation a preformed, nonstructural insulating board of foamed plastic or cellular glass; rigid boards may be applied over a roof deck, over wall framing as sheathing, in cavity walls, or beneath an interior finish material; the foamed plastics are flammable and give off toxic fumes when burned; some rigid foam boards may be used in contact with the earth and are impervious to moisture

Rigid Frame similar to an arch; a beam or column assembly that resists both lateral and vertical forces due to its rigid connecting joint; often made of laminated wood, steel, or reinforced concrete, rigid frames are used to create large, unobstructed floor areas; steel rigid frames typically form one-story structures used for light-industrial buildings, warehouses, and recreational facilities; they are often left exposed in unprotected noncombustible construction; steel is used to span from 30 to 120 feet up to 30 feet in height with corrugated metal roofing and siding; connections are bolted or welded to resist forces **(Figures R-7a and b)**

Rigid Joint a structural connection in which members cannot rotate; stress may be transferred from member to member; also called *fixed connection, fixed joint,* or *rigid connection*

Rigid Nonmetallic Conduit see *Conduit*

Rim Joist a joist that sits parallel to the edge of a sill plate, secured to the joists running perpendicular to it to provide stability to those ends **(Figure R-8)** (see also *Edge Joist*)

Rise the vertical distance between two points

Riser the vertical distance from the top of a tread to the top of the next higher tread; in stairs, an open riser is the air space between a tread projecting beyond the face of the riser immediately below

Risk Analysis an evaluation of the cost versus the benefits obtained

Rivet a metal pin used for fastening two or more pieces of metal

Rolled Members a one-piece steel structural member

Romex a trade name for nonmetallic sheathed electrical cable

Roof the external upper covering of a building, including the frame for supporting the roofing, designed to shelter the interior spaces of a building from the elements; as in floor systems, a roof must be structured to span across space and carry its own weight, the weight of any attached equipment, and accumulated rain and snow; because the gravity loads for a building originate with the roof system, its structural layout must correspond to that of the column and bearing wall system through which its loads are transferred down to the foundation system; roofs are generally weaker than floors, because they are designed to support smaller live loads; roofs are subject to wear and deterioration due to time and exposure to the elements, and they are often renovated, altered, and repaired; additional loads may be added to roofs for which they were not designed; many types of roof construction create concealed spaces/voids that allow for fire extension as well as challenges for vertical ventilation **(Figure R-9)**

- ***Flat roofs*** are found on all types of buildings, including large-area warehouses, factories, shopping centers, schools, mercantile and industrial buildings, apartment complexes, and high-rises; flat roofs have a slight slope (minimum 1/4 inch per foot up to 2 inches per foot) with interior drains and perimeter scuppers for overflow drains; the structure of a flat roof may consist of reinforced concrete slabs, flat timber or steel trusses, timber or steel beams, and decking or wood or steel joists and sheathing

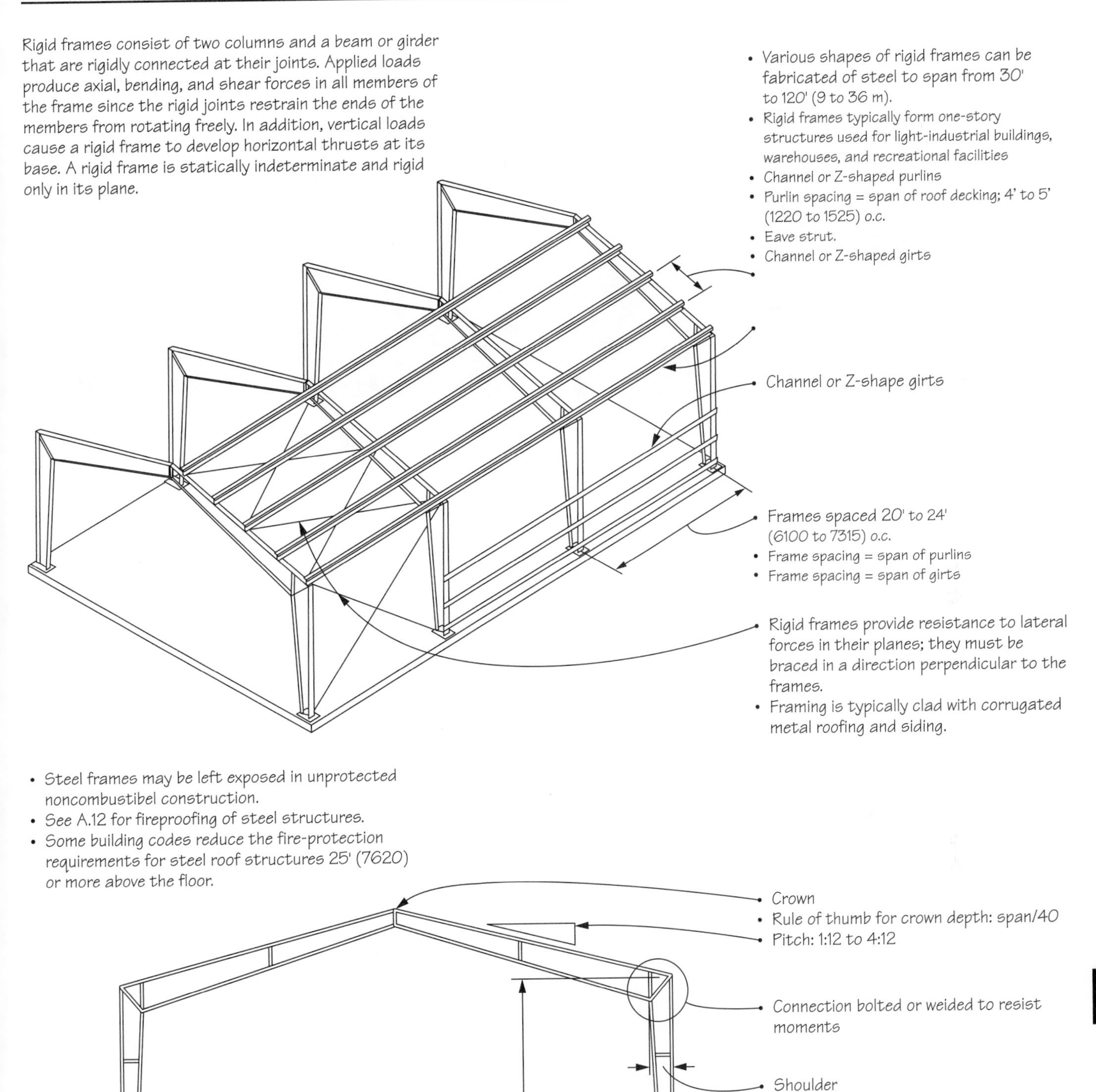

Figure R-7a Rigid Frame. *(Courtesy of* Building Construction Illustrated, *Third Edition, by* Francis D.K. Ching, *© 2008, Reprinted with permission of John Wiley & Sons, Inc.)*

Figure R-7b Rigid Frame.

Figure R-8 Rim Joist.

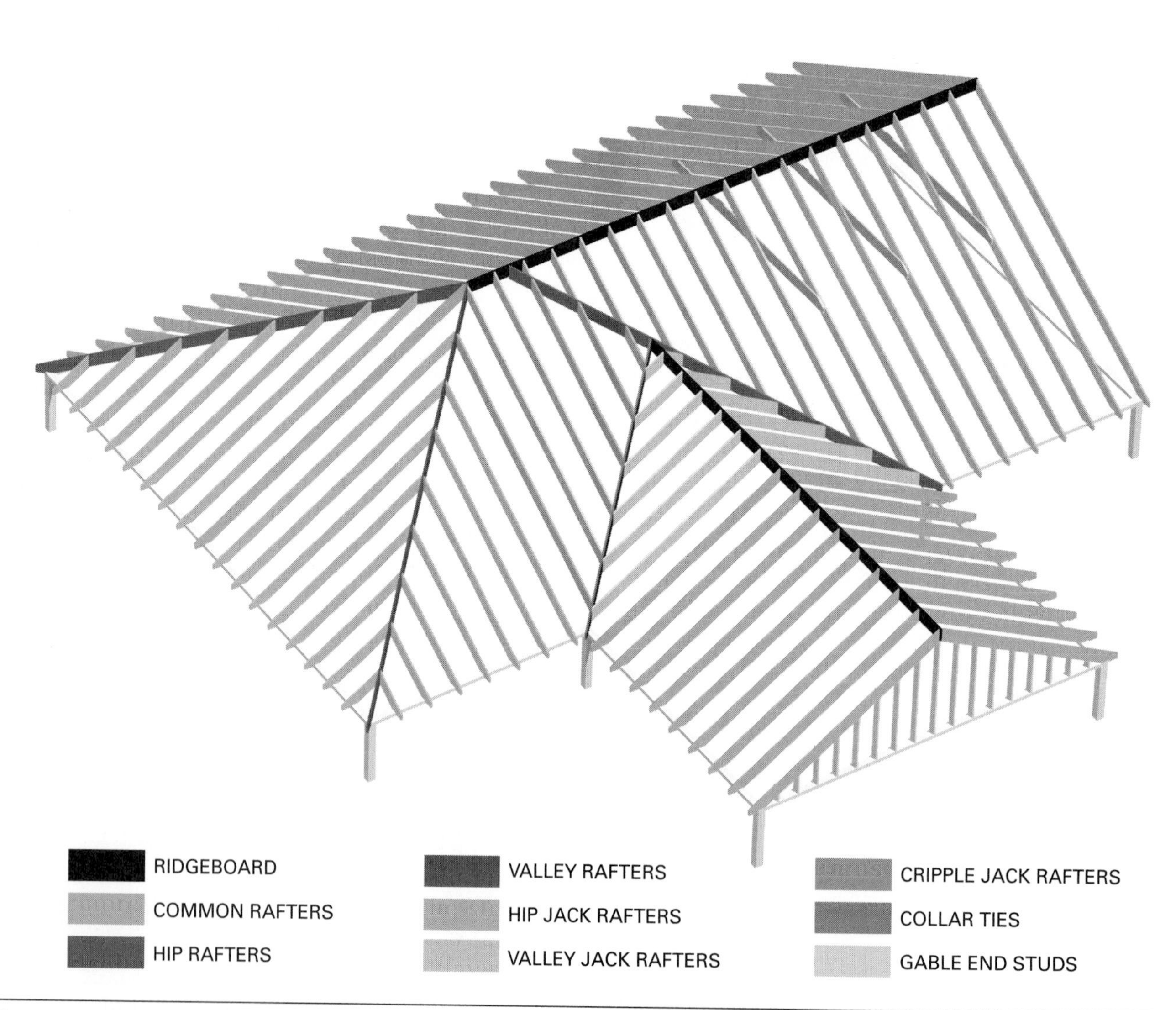

Figure R-9 Roof.

- ***Reinforced concrete roof slabs*** serve as supporting members and as the roof decking; these slabs may be formed and cast on site in the same manner as concrete floor systems or may be brought in as precast concrete slab units and topped with concrete; precast slab units must be tied together with steel reinforcement to create a horizontal diaphragm and transfer lateral forces to shear walls; roof slabs may be supported by reinforced concrete columns, reinforced concrete frames, or bearing walls of reinforced concrete or masonry; a typical roof covering for roof slabs would be a combination of a vapor retardant, rigid foam or lightweight concrete insulation, roofing membrane, a drainage layer, and a wear course; the roofing membrane is the waterproofing layer of the roof
- ***Built-up roofs*** have several overlapping layers or roofing felt saturated with tar or asphalt; four layers of 3-foot rolls of felt are typical (see also *Built Up Roofing* **(Figure B-10)**)
- ***Single-ply Roofs*** single-ply membranes may be applied in liquid or sheet form; neoprene, rubber, and PVC are thin, flexible, and strong single-ply materials that are adhered to the flat roof surface; a drainage layer permits rain water to flow to the roof drains, and the wear course protects the roofing with aggregate or pavers
- ***Flat timber trusses*** or ***steel trusses*** have parallel top and bottom chords with triangulated web members; span ranges for flat trusses are typically 40 to 110 feet; roof decks may be sheathing, roof planks, or slabs; in multistory buildings with flat roofs, it is typical for the roof to be constructed using the same structural system as the floors; typical roof coverings for flat timber or steel trusses would be built-up and single-ply roofing systems; timber or steel beams and decking create flat roof structures similar to the way timber or steel floors are framed; girders support roof beams that may support purlins and steel or wood decking. Typical roof coverings are built-up and single-ply systems
- ***Wood or steel joists and sheathing*** for flat roofs are similar to that for floor-joist framing; wooden roof joists, both solid-sawn 2-by lumber and I-joists, and open-web steel joists may bear on the wall or beam or may be attached to an anchor in the wall; roof decking may consist of metal, plywood panels, or cementitious roof planks; a roofing membrane over rigid foam or lightweight concrete insulation is typical
- ***Sloping roofs*** have inclined surfaces. They may be either low-slope roofs, up to 3:12, or medium- to high-slope roofs, 4:12 to 12:12; sloping roofs have a structure of wood or steel rafters and sheathing, timber or steel beams, purlins and decking, or timber/steel trusses; sloping roofs shed water to eave gutters; roof forms or types include gable, hip, gambrel, mansard, butterfly, monitor, sawtooth, and arched; the roof slope affects the choice of roofing material: Low-slope roofs require rolled or continuous membrane roofing, and medium- and high-slope roofs may be covered with shingles (fiberglass, asphalt, wood), tiles, or sheet materials
- ***Wood or steel rafters and sheathing*** may be constructed with light-gage steel members or solid-sawn 2-by lumber or I-joists in light wood frame construction; rafters are typically spaced on 12, 16, or 24-inch centers, depending upon the roof loads and the spanning capability of the roof sheathing; sheathing over wood or light-gage metal rafters typically consists of rated plywood
- ***Timber or steel beams, purlins, and decking*** may be supported by timber, steel, or concrete columns; timber or steel girders; or a reinforced concrete or masonry bearing wall; wooden plank-and-beam roof structures can be framed a variety of ways, depending upon the direction and spacing of the roof beams, the elements used to span the beam spacing, and the overall depth of the construction assembly; rafters, roof beams, purlins, and roof decking may be used parallel or perpendicular to the slope; coverings include fiberglass, asphalt and wood shingles, wood shakes, tiles, and corrugated metal roofing
- ***Timber or steel trusses*** are typically prefabricated and used to span areas 40 to 150 feet in length; inclined truss types include Pratt, Howe, Belgian, Fink, Scissors, and Bowstring; steel truss members are bolted or welded with gusset-plate connectors; purlins may support metal or cementitious roof decking; noncombustible steel trusses may be left exposed if at least 20 feet above the finished floor; wood trusses may be spaced up to 8 feet on center, depending upon the spanning capability of the roof decking or planking; when purlins span across the trusses, the truss spacing may be increased up to 20 feet; layering up to five members, and joining them at the panel points with split-ring connectors, can assemble heavier wood trusses; such wood trusses are capable to carrying greater loads than trussed rafters and are spaced further apart; wood trussed rafters are

preengineered and shop-fabricated monoplanar trusses, typically of 2 × 4 construction; typical spacing is on 2-foot centers but may vary to up to 4 feet; truss members are connected with metal-toothed plate connectors, and trussed rafters typically span from 20 to 32 inches; pitch range is 2:12 to 8:12; sheathing requirements are similar to those for conventional rafter framing, and roof coverings include the same matieral used for other sloped roofs

Roofing also called *roof coverings,* the materials used on the exposed surface of the roof, which may depend upon the slope, condition, and desired look; materials include rolled or continuous membrane roofing for slopes up to 3:12 and shingle or sheet roofing for slopes 3:12 and higher; Asphalt-saturated felt is typically used for underlayment to protect the roof sheathing; Roofing materials for sloped roofs include composition shingles, wood shingles and shakes, rolled roofing, metal roofing, slate, and clay or concrete tile.

Roof Assembly the components(s) above the roof structural framing, including the roof deck, vapor barrier, insulation roof cover, coatings or toppings or any combination there of; the system consists of a roof covering and roof deck or a single component serving as both the roof covering and the roof deck

Roof Coverings the materials used on the exposed surface of the roof; dependent upon slope and conditions, typical roof coverings include rolled or continuous roofing membrane for slopes up to 3 and 12 and shingle or sheet roofing for 3:12 slopes and higher

Roof-covering Classifications includes A, B, C, and nonrated categories

Roof Decking plywood, hardboard panels, or space boards laid on top of the rafters or other roof support framing onto which the roof coverings, such as shingles, are fastened

Roof Drain a drain installed at the low spot on a roof that prevents water from standing on a flat or nearly flat roof; the water drains into a drainpipe and is carried to the ground or into a drainage system

Roof Drain Strainer a strainer at the inlet to a roof drain designed to trap debris and prevent it from entering and clogging the drainpipe

Roof Failure examples of line-of-duty deaths due to roof failure include the following:

- **Cliffside Park, NJ 1967** Five firefighters perished after a roof collapse
- **New York, NY 1978** Six firefighters died due to failure of a bowstring timber truss in a supermarket
- **Orange County, FL 1989** Two firefighters died in a commercial structure
- **Hackensack, NJ 1988** Five firefighters died in a car dealership due to a bowstring timber truss failure
- **Lake Worth, TX 1999** Three firefighters died in a church when a lightweight truss failed

Roof Jack a specialized piece of flashing with a sleeved hole that fits over a vent stack that protrudes through a roof; the flashing seals the opening and diverts rain around the vent stack, preventing leaks

Roof Pitch see *Roof Slope*

Roof Rise the height or vertical rise of a roof measures from the wall plate, on which the rafters rest, to the peak of the roof ridge

Roof Run the horizontal distance covered by the roof from the top plate of the wall to the midpoint, equal to one-half the span of the roof

Roof Sheathing the structural covering of the rafters or trusses, usually plywood or hardboard panels or closely spaced boards; the roof covering is attached to the sheathing

Roof Slope the angle of a roof expressed as the rise over the horizontal distance; a four-and-twelve slope (written *4:12* or *4/12*) equates to 4 inches of rise for every 12 inches of horizontal distance

Roof Structure structure above the roof or any part of a building enclosing a stairway, tank, elevator machinery, service equipment, or part of a shaft extending above the roof

Roof Systems the primary sheltering element protecting the interior space of the building from the elements, engineered to carry its own weight as well as live loads such as snow, ice, and wind; design loads for roofs are less than those for floors, but roof systems may also accommodate mechanical and electrical equipment; the roof form, spacing, span, and slope affect the choice of finish roofing material, the interior ceiling system, and the building's interior layout; structural roof systems include joists/rafters and sheathing, plank and beam, flat and pitched trusses, and concrete panels

Roof Tile tiles of clay, cement, or other material designed for use as a roof covering; may be attached to the roof deck with nails or wire

Roof Truss a rigid framework, made up of triangular sections, that spans across the walls of a building and which is used, in combination with other trusses, to support the roof; roof trusses are usually prefabricated and delivered to the building site ready to install; this is especially true for large

Figure R-10 Roof Ventilation.

jobs involving several structures; roof truss materials include wood, steel, and combinations of both (see also *Wood Trussed Rafters* **(Figure W-10)**)

Roof Ventilation the natural or mechanical process of supporting air to or removing such air from attics, cathedral ceilings, or other enclosed spaces over which a roof assembly is installed; firefighters perform roof ventilation to assist with the rapid removal of heat and the products of combustion **(Figure R-10)**

Rotate to turn about an axis

Roughing-In the installation of a system, such as electrical or plumbing, that will be behind finished surfaces and not shown when the structure is complete

Rough Lumber rough-sawn, full-dimension lumber with an uneven, unfinished surface on all four sides; also called *rough-hewn lumber*

Rough Opening a rough-framed opening in a building wall intended for the installation of a window, door, or other such device

Row Houses one of several houses built adjacent to one another without open space between sidewalls; intervening walls may be party walls, but in some instances, each house is self-supporting and is separated from its neighbors by unpierced fire walls

Rubble Wall a wall composed of an inner and outer wythe of coursed masonry, with space between filled with random masonry, sometimes mixed with mortar; these walls, often unreinforced, are unstable during earthquakes and perform poorly when beams or trusses fail

Run the horizontal distance covered by a rafter or flight of stairs: in roofs, the horizontal distance runs from the eaves to the ridge of a sloping roof; in stairs, it is the horizontal measurement of a stair tread or the distance of the entire length

Running Bond a bricklaying pattern of all stretcher courses, each course offset from the next by half a brick or sometimes a third, depending on the style of running bond

Saddle a ridge connecting two higher elevations of a roof (see also *Cricket*)

Safety Factor ratio of the strength of the material just before failure to the safe working stress

Sally Port a security vestibule with two or more doors where the intended purpose is to prevent continuous and unobstructed passage by allowing the release of only one door at a time

⚠ **Sand-lime Mortar** a water-soluble mortar mix made only with sand and lime; used in the 1800s and early 1900s in unreinforced masonry construction, this mortar tends to deteriorate over time and may not bind well to masonry units and was replaced by Portland cement for its superior adhesion and durability

Sash a movable window frame containing one or more panes of glass

Saturated Felt felt that has been impregnated with an asphalt compound to make it water-resistant; it is used under shingles or other roof coverings as part of the moisture barrier in construction applications where water resistance is needed, such as behind a shower wall

Scaffold a temporary wood or metal structure erected in or around a building to permit access to work being done above the first floor level during construction or for maintenance purposes; also called *scaffolding* or *staging*; hanging scaffolds are suspended and can be raised and lowered, as on high-rise buildings when exterior work or maintenance must be done

Scissors Roof Truss a roof truss with a steep exterior slope and a bottom chord that angles up in the middle, providing sloped interior ceiling surfaces (see also *Wood Trussed Rafters* **(Figure W-8b)**)

Scratch Coat the first surface coat of plaster or stucco, which is roughened to ensure the bonding of the second coat

Screed a rigid, straight piece of wood or metal used to level concrete or other materials that have been poured into forms; also used to remove the excess material

Screen a fine mesh, usually wire or nylon, used as a covering, barrier, or filter; frame-mounted screen material is commonly used over windows, doors, and other openings in buildings to keep insects and debris out

Screw a fastener with a raised spiral ridge down the shank and a flared, slotted head slightly larger in diameter than the shank

Scupper a drain on a roof or other deck, or through the side of a building, to allow rainwater to run off; by identifying scupper locations, firefighters are able to determine the roof height hidden by the wall

Scuttle a small opening in a roof or ceiling that permits access to an attic or rooftop

Sealant a caulking compound designed to form an airtight or waterproof bond at a joint while remaining flexible

Secondary Collapse a collapse following the initial collapse (See also *Collapse*)

Section View building plan that shows a cross section of the building, as if it were sliced to reveal its skeleton

Self-Closing equipped with an approved device that ensures closing after opening

⚠ **Self-releasing Floors** in some heavy timber buildings, floor girders are set on brackets attached to columns. A loose connector (cleat, dog iron) allows the floor to collapse in order to prevent masonry walls from collapsing. An older construction technique for floor beams called *fire cut* also allowed the beams to fall away from the walls, which allowed for more rapid rebuilding after fires (see also *Fire Cut*)

Sequence an order of events or tasks; sequencing tasks such as foundation layout and pour, wall framing, roof framing, rough plumbing and electrical, and so on are an important step in scheduling and ordering for building construction; officers at structure fires do mental sequencing to determine appropriate course of actions that may involve simultaneous tasks

Service Entrance the interface of utility lines with a building's wiring. The utility service may come from an overhead service drop or from an underground cable connection

Service Entrance Cable factory-assembled electrical wiring consisting of two or three insulated conductors with a stranded, uninstalled neutral conductor wrapped around them

Service Lateral underground power lines from the power line transformer to the meter box in a building

Service Mast the vertical section of conduit through which the service wires to a building pass before reaching the service entrance breaker box

Service Panel an electrical panel containing circuit breakers that serves as a distribution point for the electrical service entering the building

Shaft an enclosed space extending through one or more stories of a building, connecting vertical openings in successive floors and sometimes the roof; used to accommodate elevators, dumbwaiters, mechanical equipment, or similar devices or to transmit light or ventilation air

Shake Roof a roof covered with wood shake shingles; shakes are longer, thicker shingles

Shear a force or stress that results when two forces act on a body in opposite directions in parallel adjacent planes: a force that slides two surfaces past each other

Shear Force a force acting perpendicular to a plane; for example, the force of gravity acting on a wall cabinet is perpendicular to the axis of the screws holding the cabinet in place

Shear Panel a floor, wall, or roof designed to serve as a deep beam to assist in stabilizing a building against deformation by lateral forces

Shear Plate a plate incorporated into a structure that is designed to resist shear; a plywood panel mounted on a wall from the bottom plate to the top plate with all the edges blocked is a shear plate

Shear Wall a wood, concrete, or masonry wall that can resist changes in shape and can transfer lateral loads to the ground through the foundation; sometimes referred to as a *vertical diaphragm*; shear walls are designed to resist horizontal loads and provide lateral stability by using rigid joints, diagonal bracing, or sheathing applied to the exterior or interior walls; in wood frame construction, this is done by mounting plywood or OSB sheathing from the bottom plate to the top plate and nailing it in place according to a shear schedule provided in the plans; the sheathing adds considerable stiffness to a stud wall, giving it the ability to resist sideways or horizontal loads

Sheathing boards or sheet materials that are applied to the framing of a building to which siding, flooring, or roofing are applied; shear plywood is nailed to the outside walls to stiffen the structure and provide nailing for exterior finish materials; structural sheathing is capable of bracing the plane of a framed wall or roof, and diagonal sheathing uses boards applied diagonally for lateral strength. Sheathing may be tongue-and-groove (T&G), plywood, OSB, or gypsum wallboard **(Figure S-1)**

Figure S-1 Sheathing. (*Courtesy of Kathleen Siegel.*)

Sheathing Paper a moisture-resistant building paper that is applied to roofs or walls as a weather barrier before the final covering is installed

Shed a small structure for storage or other purpose

Shed Dormer a dormer with a shed roof

Shed Roof a roof style in which the slope from the peak to the eave line is in one direction only

Sheet Metal metal that is in the form of thin sheets; it has several uses in building, such as in the manufacture of flashing, vents, and ductwork

Sheet Piling pilings made of wide sheets of corrugated steel or a combination of metal sheets and lumber; they are used to hold back earth at the sides of an excavation

Shell a thin, curved plate, often of concrete, used as a roof

Shells also known as *domes*

Shim to use thin, tapered pieces of material, such as wood, to level or plumb a structure or part of a structure or surface

Shingle a unit of material intended for use as a weatherproof roof covering or siding; shingles come in a variety of shapes and styles and are made of several different materials, such as asphalt, fiberglass, cedar, slate, clay, and metal

Shiplap Siding a type of horizontal siding that uses boards joined edge to edge with overlapping rabbet joints; other horizontal siding types include bevel, drop, and Dolly Varden (see also *Siding*)

Shoring temporary vertical or horizontal supports

Shoring Jacks jacks that exert compressive force on the shoring installed along trench walls to prevent a cave-in; the jack holds the shoring in place by putting pressure on trench walls (see also *Reshores*)

Short Circuit an unintentional or accidental failure in an electrical circuit in which a portion of the current is diverted to a conductor that is not a normal part of the ground or point on the circuit

Shotcrete pneumatically applied concrete; air pressure forces the concrete through a nozzle, a method that permits the spraying of concrete on walls and curved surfaces

Shutters a covering for a window, usually louvered for ventilation, mounted on hinges on each side of the window

Siding a weatherproof material, such as shingles, boards, or units of sheet metal used for surfacing the exterior walls of a frame building; may be made of wood, asphalt vinyl, aluminum, or other weatherproof materials; bevel siding is wedge-shaped and is used by lapping each board over the edge of the board below; shiplap siding is rabbeted at the top so that each piece of siding overlaps and fits snugly against the adjoining piece; other examples include board-and-batten siding (vertical boards with strips nailed over the joints), wedge-shaped clapboard, plywood siding (4-foot sheets, side-matched, grooved, "T111"), shingles and shakes (wood), asbestos cement shingles, asphalt felt siding (which may look like brick or stone but is combustible), vinyl siding made to look like wood (also combustible), metal siding made to look like clapboard or imitation stone, corrugated metal siding (industrial building applications typically), stucco (a thin concrete surface over wood, brick, or masonry), and brick veneer

Sill the lowest horizontal member of a frame structure, resting on, or anchored to, a foundation wall, also called a *mudsill* or *sill plate*; may also refer to the bottom exterior member of a window or door or the masonry below

Simple Cornice a frieze board or horizontal board placed flat against a wall to form a cornice

Simply Supported Beam a beam supported at both ends

Single Coursing the application of siding shingles in a single layer with a small amount of overlap along the edge of the course below

Single-Phase alternating current electricity with one phase typically found in residential occupancies; electricity is generated in three phases of power, 120 degrees apart

Single-ply Roof roof covering made up of a layer or layers of insulation and a single sheet of membrane material used for waterproofing; the membrane may be applied in liquid or sheet form, and materials include thermoplastic membranes, polyvinyl chloride (PVC), neoprene, butyl rubber, polyurethane, and polymer-modified bitumen; these materials are very thin, flexible, and strong; some are reinforced with fibers and may have coatings or gravel ballast (see also *Membrane Roofs*)

Single-ply Membrane a roofing membrane applied using one layer of membrane material, either homogeneous or composite, rather than multiple layers

Single-wall Siding a building exterior wall that has a single siding covering, as opposed to siding over sheathing; most siding can be installed in this manner

Site the physical location of a construction job or the location of proposed construction; the parcel of land bounded by a property line or a designated portion of a public right-or-way

Skeleton the framework of a building or other structure

Skeleton Frame a system of columns and beams that supports walls **(Figure S-2)**

Figure S-2 Skeleton Frame.

S

Sketch a rough drawing

Skew a turning out of alignment or casting at an angle

Skewed out of square with relation to a standard reference; not parallel, not level, not plumb; at an angle

Skewed Beam a structural member that intersects another member at something other than a 90-degree angle

Skim Coat a thin coat of drywall compound

Skin an exterior layer or surface coating

Skirt siding at the base of a structure, such as a mobile home or porch, which extends from the bottom of the structure to near the ground

Skirting baseboard around the base of an interior wall

Skylight A roof opening covered with glass or plastic designed to let in light and sometimes ventilation to the top floor; may help to identify hallways in multiple family dwellings **(Figure S-3)**

Figure S-3 Skylight.

Slab any thin, flat element, usually a floor or roof. Typically thought of as a concrete floor, either poured in place or precast. A concrete slab may be placed at or near grade level to serve as a combined floor and foundation system. Precast concrete slabs are used above ground for greater structural efficiency, which results in less depth, reduced weight and longer spans

Slab-and-Beam Frame a construction technique using concrete slabs supported by concrete columns

Slab-jacking forcing cement under a cured concrete slab to raise the height of the slab

Slag Concrete lightweight concrete in which furnace slag is used for aggregate

Slake hydrating or adding water to a material, such as lime, to cause it to disintegrate or crumble so that it can be used in mixtures

Slate a fine-grained rock that easily splits along thin planes to form sheets; used as a roofing material, for decorative flooring, or as an exterior wall facade

Sleepers wood members embedded in poured concrete to be used as a nailing surface for wood framing or flooring

Sleeve a hollow cylinder through which another commodity can pass, such as the metal sleeve installed in poured concrete to allow for the passage of piping or conduit

Slenderness Ratio the ratio of the length of a structural member to the thickness of the member; provides information on the stiffness of a member, as stiffness is proportional to length and thickness

Slip Form a concrete form for tall structures that is raised slowly as vertical concrete pours are made and set up; the form is raised by means of a jacking mechanism and placed into position for the next pour, which is made on top of the portion of the wall or column hat has already set, extending it

Slope the angle of deviation from level; the relative steepness, calculated by the horizontal distance and elevation change between two points; a slope may be expressed in degrees of angle, as a ratio or horizontal distance to vertical drop, as a decimal, or as a percent of vertical change to horizontal distance (see also *Pitch*)

Sloping widening a trench across the top so that it has sloping sides to minimize the risk of a cave-in

Slumps a test to measure concrete consistency by measuring the amount that a quantity of concrete of specified shape and size will slump due to the effects of gravity when unsupported on the sides; also refers to the stiffness of a mix of concrete

Slurry a mixture of liquid and solid particles in which the particles are suspended in the liquid; pumped concrete is an example of such a mixture

Smoke Barrier a continuous membrane, either vertical or horizontal—such as a wall, floor, or ceiling assembly—that is designed and constructed to restrict the movement of smoke

Smoke Chamber the open area above the combustion chamber in a fireplace assembly that traps the smoke and, with the aid of the smoke shelf and damper, prevents downdrafts; the smoke chamber forms the support for the chimney flue, which rises above it

Smoke Damper a listed device installed in ducts and air-transfer openings, designed to resist the passage of smoke; usually controlled by a smoke-detection system

Smoke Exhaust System a mechanical or gravity system intended to move smoke from the smoke zone to the exterior of the building

Smoke-Proof Enclosure consists of an enclosed interior exit stairway designed to limit the movement of products of combustion produced by a fire

Soffit a lower horizontal surface, such as the undersurface of eaves or cornices, or the false spaces above cabinets **(Figure S-4)**. See Steel Joist Framing on page 101 and Structural Failure on page 104. (see also *Exterior Soffits, Interior Soffits* **(Figure I-3)**)

Figure S-4 Soffit.

Soffit Vent ventilation holes or openings in the soffit under the eaves of a building

Soft Story in a multistory building, a lower story that has less lateral stiffness or strength than the stories above it; the lower story is often used for entryways, lobbies, or vehicle parking areas **(Figure S-5)**

Sole Plate the horizontal board to which the bottoms of wall studs and other framing members are attached; also called the *bottom plate, sole,* or *sole piece*

Solid Bridging solid blocks of wood placed between floor joists to distribute concentrated loads

Solid-core Door a door made of solid pieces of material with no voids in the door structure

Solid Masonry Wall a masonry wall made of units that are three-quarters solid material with mortar at all joints

Sonotube fiber cylinders coated with wax and used as concrete forms, typically 6 to 48 inches in diameter

Soot Pocket a space at the bottom of a chimney flue where soot can accumulate and be removed through a clean-out door

Space a definable open area, such as a room, hall, assembly area, lobby, or courtyard

Space Frame a three-dimensional truss

Spall flaking or chipping of a concrete surface caused by weathering, expansion, or a blow

Spalling deterioration of concrete through the loss of surface material due to the expansion of moisture when exposed to heat

Span the distance between structural supports; the spanning capability of horizontal elements determines the spacing of their vertical supports

Span, Allowable the maximum permissible distance that a structural member may span between supports

Span Ranges the distance range between two points that will vary based upon a structural element's size and strength; for timber planks and joists, spans range from 15 to 20 feet; for laminated timber beams, the span is up to 80 feet; timber trusses span approximately 100 feet; steel decking about 16 feet; steel wide-flange beams up to 60 feet; and open-web steel bar joists up to 90 feet; for concrete, reinforced slabs can span about 20 feet, joist slabs 36 feet, precast planks 40 feet, and precast reinforced concrete tees can span over 100 feet

Spandrel an area between and above two connected arches, or between one arch and the ceiling; in a steel frame building, the area between the top of a window and the sill of the window directly above it

Spandrel Girder a beam running from column to column in outside walls that carries the curtain wall above it

Figure S-5 Soft Story.

Spandrel Space the distance between the top of a window and the bottom of the window above it

Spandrel Wall the section of wall spanning the area from the top of one window to the sill of a window directly above in the next story of the building

Spark Arrestor a screen or expanded metal covering on the outlet of an exhaust or a chimney that allows smoke to pass through but prevents sparks from exiting and creating a fire hazard

Spray Polyurethane Foam Roofs a roof material that may be applied directly to an existing built-up roofing system

Sprayed Fire Resistance a cementitious or fibrous material that is sprayed-applied to structural elements, walls, floors, and roofs to provide fire-resistive protection

Splice to join two pieces together

Splined Joint a butt joint in which a groove is cut in each of the members being joined and a spline is fitted into the adjoining grooves to add strength to the joint

Split-Faced Block a masonry building block with a rough textured face on one side

Split-Level House a house with floors at different levels less than one floor apart

Split-Ring Connector a type of timber connector that consists of a metal ring inserted into corresponding grooves cut into the faces of the joining members and held in place by a single bolt; used in wood post-to-beam connections and timber trusses, split-ring connectors enlarge the area over which a load is distributed

Spray-on Fireproofing a mixture of mineral fibers and an inorganic binder applied by air pressure with a spray gun to provide a thermal barrier to the heat of a fire **(Figure S-6)**

Spreader any member used to distribute loads or forces over a wide area; in concrete formwork, the wood or metal spacer used to properly distance the forms

Spreaders steel tie rods and anchors (star plates, circles, channels) used to support walls typically found in Type III unreinforced masonry buildings; may connect two opposite walls together or floors and roof joists together; unprotected steel tie rods conduct heat in a fire and may lose their ability to keep walls plumb **(Figure S-7)**

Figure S-6 Spray-on Fireproofing.

Figure S-7 Spreaders.

Sprinkler a water-pipe fitting that disperses water in a spray; fire sprinklers typically operate with heat-activated heads

Square a four-sided geometric figure, with all sides equal and all angles 90 degrees

Square Butt Strip Shingle a commonly used asphalt roofing shingle that is 3 feet long and a foot wide; it has three tabs, formed by two slots cut into the shingle on the long edge that extend partway across the shingle width

Stack a chimney-laying pattern in which each course of stretchers is laid directly in line with the adjacent courses, and all vertical joints are aligned

Stack Bond a bricklaying pattern in which each course of stretchers is laid directly in line with the adjacent courses, and all vertical joints are aligned

Stage a space within a building used for entertainment and utilizing drops or scenery or other effects

Staggered-stud Partition a wall structure with plates of wider material than the stud materials, in which the studs are alternated from one side of the plate to the other; also called *offset studs*

Stairways used to connect floors and lofts, code requirements specify width, rise and run, and landings; types include spiral, L, U, and straight run

Standard Brick a brick with dimensions of 3-5/8 × 8 inches

Standing-seam Metal Roof a roof covering made of metal panels joined with interlocking seams that fold over at right angles to the plane of the roof panels

Standpipe a vertical run of pipe used as a water supply to a fire-protection system

Standpipe System an arrangement of piping, valves, and hose connections installed in a structure for the purpose of extinguishing a fire; standpipes are connected to a water supply system or by means of pumps, tanks, and other equipment necessary to provide an adequate supply of water to the hose connections

Starter Strip a strip of wood nailed on the outside of the framing against which the siding will rest or the first row of roofing at the eave line in the application of asphalt shingles; in roofing, the starter strip is laid, and the first course of shingles is laid on top of the starter strip

Static showing little change or movement

Static Loads loads that are steady, constant, or gradually applied; dead loads are static loads

Stay-in-Place Forms concrete forms designed to remain in place as a permanent part of the building structure

Steel a strong alloy of iron, carbon, and other metals that combines high strength (tension and compression) and stiffness with elasticity; probably the strongest low-cost material available, steel is considered by some to be the ideal structural material: it includes high strength per unit weight and high strength per unit volume compared to reinforced concrete and masonry structural elements; steel construction has allowed structural frames to become taller and thinner; used for structural members such as beams, girders, large trusses, lintels, and columns, steel members are joined together by rivets, welds, and bolts; steel conducts heat and requires insulation from fire to prevent distortion under high temperatures, as it loses its strength under high-heat conditions (over 1000 degrees F) it will elongate or warp and may push through any barrier or load-bearing member; corrosion protection and maintenance is also required; steel left exposed to moisture and oxygen will turn to rust (iron oxide), which can lead to a loss in structural integrity

Steel Columns transmit gravity and lateral loads down to the foundation system; the most frequently used section for steel columns is the wide-flange (W) shape; it is suitable for connections to beams in two directions, and all of its surfaces are accessible for making bolted or welded connections; other steel shapes used for columns are round pipe and square or rectangular tubing

Steel Construction construction in which the main structural members, including studs and rafters (where applicable), are made of shaped structural steel; the structure can be very strong while maintaining a degree of flexibility that provides earthquake resistance **(Figure S-8)**

Figure S-8 Steel Construction.

Steel Floor Decks manufactured from copper-alloy sheet steel, decks may be either corrugated or cellular in form, which provides a permanent formwork and steel reinforcement for structural concrete to form a composite floor slab; specific forms, patterns, widths, lengths, gages, and finishes vary by manufacturer, and spans vary from 2 to 16 feet with slab thickness from 2-1/2 to 5 inches

Steel Floor Systems includes steel joint systems and steel beams and decking

Steel Frame Construction building construction in which the frame consists of columns that support a grid of girders, beams, and open-web joists; column spacing equals the beam or girder spans exterior walls are typically non load-bearing curtain walls supported by columns alone or by columns, beams, and edges of floor slabs steel beams and columns carry vertical loads; lateral-force–resisting systems include brace frames, moment-resisting frames, prefab metal buildings, frames with URM infill walls, and frames with concrete cast-in-place or reinforced masonry walls. Steel column shapes include wide-flange (W), round pipe, and square or rectangular tubing **(Figure S-9)**

Steel Joist Framing (Open-web Bar Joists) relatively lightweight construction for floor and roof systems; typically shop fabricated as steel angles or rods in standardized lengths, depths, and carrying capacities; types include standard (J and H series) 8- to 30-inch depths spanning up to 60 feet; long span (LJ or LH series) have 18- to 48-inch depths spanning up to 96 feet; and deep longspan (DLJ and DLH series) have 52- to 72-inch depths spanning up to 144 feet; spans are typically supported on masonry walls, concrete walls, or steel beams; decking types for steel joist framing include cast-in-place reinforced concrete or gypsum slab; precast concrete or gypsum plank; composite, steel-reinforced concrete slab; or wood plank; joist spacing equals the span of roof decking, and joists may frame into a load-bearing wall, steel frame support, or it may bear on a wall

Figure S-9 Steel Frame Construction.

for a flush roof edge; cross or straight bridging is used between joists, typically on 8-foot centers, to resist rotation or buckling; open webs may permit mechanical lines to pass through (see also *Open Web Joist* **(Figure O-2)**)

Steel Joist Roof Decks the surface that forms the roof decking, which includes solid wood decking, steel decking, and cementitious roof planks

Steeple a tall, tapered tower structure on top of a building, often used in church construction to house a bell or chimes

Step Flashing flashing used where a roof slope abuts a vertical wall, skylight, or chimney, it consists of a series of short sections of flashing bent at right angles across the center, so that half of the flashing is on the roof, and half up the wall or chimney; each section of flashing overlaps the upper edge of the section below it

Stepped Footing a building perimeter foundation in which the footing descends in step-like sections, designed for use on hillsides or in changes in elevation to provide for greater economy of concrete use

Stick-Built Construction a structure built on site from individual structural members, as opposed to prefabricated sections that are delivered to the site

Stick-Built Roof a roof built on site from individual structural members, rather than with prefabricated sections, such as trusses manufactured away from the construction site and delivered for installation

Stiffness measure of the force required to push or pull a material to its elastic limit

Stock material on hand; material that is standard size or unchanged from the original manufactured condition

Stool the flat member between the jambs at the bottom of a window opening against which the bottom sash rests

Stoop a raised, usually covered porch or platform at the entrance of a house or other building

Stop a molding fastened to the inside of window jamb to secure the sash; a structural member designed to limit the travel of an object, such as a sliding door, at a predetermined point

Storm Door a door installed on the outside of an existing exterior door to provide insulation and weather protection; the door may have an interchangeable window and screen to provide ventilation during warm weather

Storm Window a window placed on the outside of an existing window to provide insulation and weather protection, usually installed for winter and removed in the spring

Story a floor or level in a building; the portion of a building located between the upper surface of a floor and the upper surface of the floor or roof next above

Straight-run Staircase a staircase leading directly from one floor to the next without turns or landings; also called a *straight flight of stairs*

Strain a deformation or change in the shape of an object caused by an applied force; internal forces that resist load, measured in fractions of an inch of deformation per inch of original length of the material

Strand one of the wires making up a tendon or cable

Strap a metal piece used to hold joints together, as in heavy timber construction and shear wall bracing

Strength the ability to carry a load, such as the ability of a load-bearing structural member to hold up the weight imposed on it without bending or breaking

Stress the action of forces on an object or system, which tend to cause deformation, strain, or separation of adjoining members. Allowable stress is the maximum permissible stress that may be placed on a structural member or a pressure-retaining component; the intensity of force per unit area is measured in PSI or KIP; three common stress modes are *tension, compressive, and shear* **(Figure S-10)**

- ***Compression*** a force pressing or squeezing a structure together; some materials have great compressive strength yet are weak in tensile strength, such as concrete
- ***Tension*** stress placed on a structural member by the forces that pull materials apart; steel has great tensile and compressive strength
- ***Shear*** causes molecules of material to slide past one another, such as a brick veneer wall breaking away from the cement bond to the back wall

Stressed Skin plywood sheathing fastened or bonded to framing in a manner that makes the combination work as a unit to resist shear stresses (see also *Diaphragm*)

Stretcher a brick laid horizontally with a narrow side of the length exposed

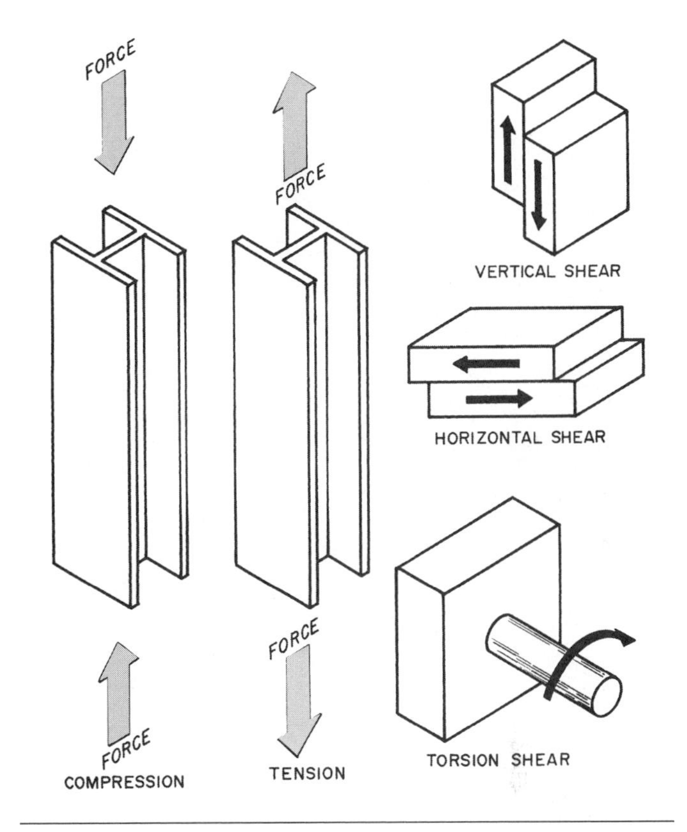

Figure S-10 Stress.

Strike Plate a metal plate, recessed flush with a doorjamb, into which a lock bolt latches

Stringer the member on each side of a stair that supports the treads and risers

Strongback a stiffener, usually a reinforcing structural member, such as a beam placed on edge **(Figure S-11)**

Structural pertaining to load-carrying parts or members; a solid part of a structure or building that is designed and installed to carry a load

Structural Attachment an attachment to an object, such as a pipe, that is designed to carry a load

Structural Block masonry building block with compression strength ranging from 1800 to 3000 psi, sufficient to be used for a wall that will carry a structural load

Structural Clay Tile fired clay, in the form of blocks that are extruded to be required shapes; tiles are durable, lightweight, fireproof, and inexpensive but have generally been replaced in modern construction by concrete block

Structural Lightweight Concrete concrete containing lightweight aggregate, used in structures designed to minimize weight

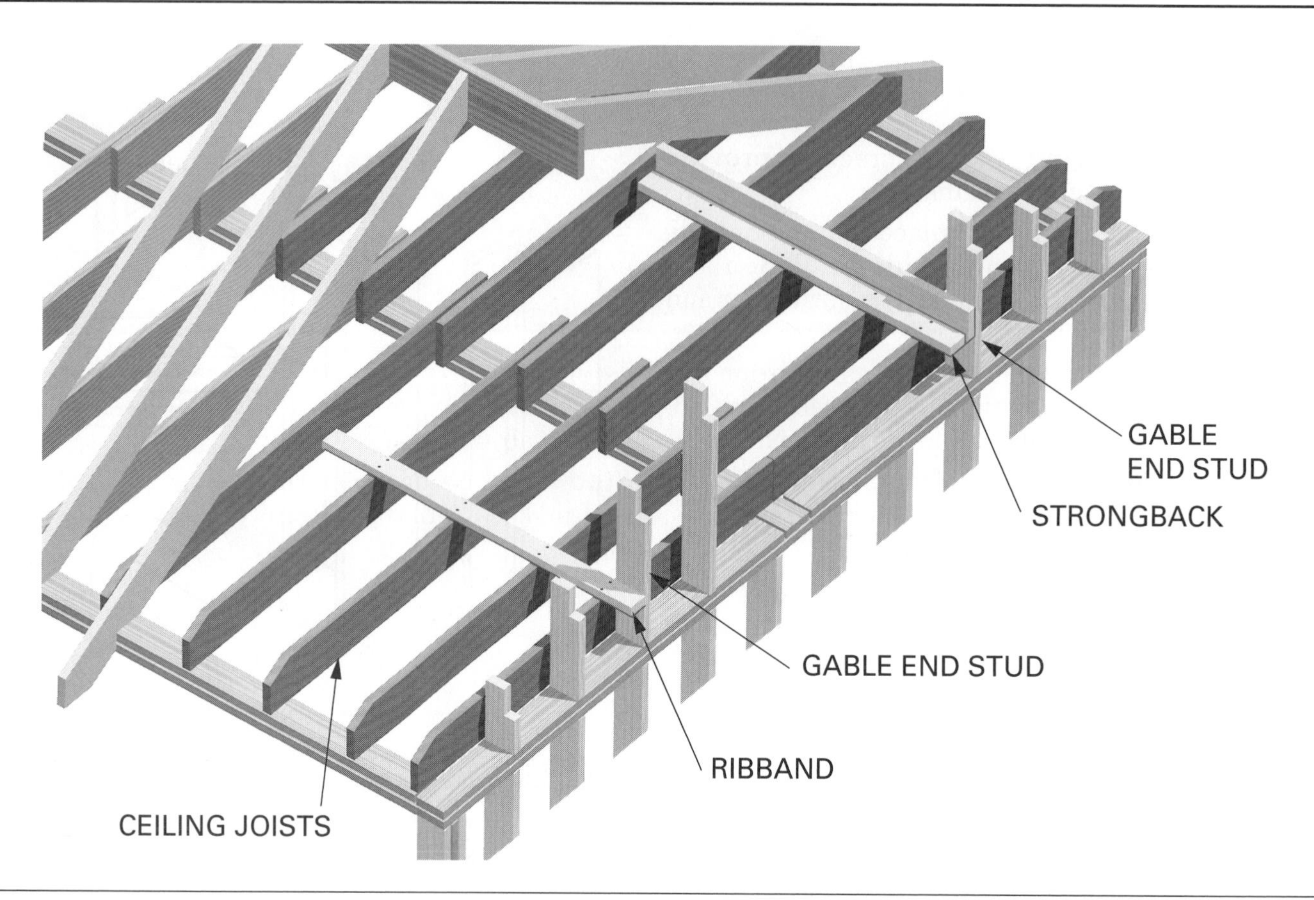

Figure S-11 Strongback.

Structural Composite Plywood Panel a plywood panel with a homogeneous, flakeboard core and tongue-and-groove edges, used in applications such as subflooring; it is more economical than regular plywood sheets of the same thickness

Structural Design the process by which structures are designed and constructed to support all expected loads without exceeding the allowable stresses of the materials employed; NFPA 5000 states that the basic requirements of strength, serviceability, self-straining forces, and analysis shall be in accordance with ASCE 7, "Minimum Design Loads for Buildings and Other Structures"

Structural Element columns, girders, beams, trusses, arches, braced frames, moment-resistant elements, and other framing members essential to the stability of the structure, designed to carry a portion of the dead or live load and lateral forces

Structural Failure any condition such as fracturing, buckling, or plastic deformation that renders a structural assembly, element, or joint incapable of sustaining the load-carrying function for which it was designed (see also *Collapse, Collapse Causes, Collapse Indicators and Failure* **(Figures C-9, C-10, C-11 and F-3)**)

Structural Fiberboard fiberboard impregnated with asphalt for water resistance, available in various thicknesses and several densities; also called *structural insulating board*

Structural Frame the columns, and the girders, beams, trusses, and spandrels having direct connections to the columns, and all other members that are essential to the stability of the building as a whole

⚠ **Structural Hazards** those parts of specific buildings that pose hazards as a result of fire or natural or man-made events; these hazards include unreinforced masonry walls, parapets, unprotected steel and lightweight engineered wood elements, facades and marquees

Structural Integrity the ability of a structure to maintain its designed load-carrying capacity; integrity includes greater redundancy and continuity in structural systems and details

Structural Joist And Plank lumber that is at least 2 inches thick and 6 inches wide or wider

Structural Light Framing conventional framing using 2 × 4 and 2 × 6 lumber, used for standard one- and two-story structures (see also *Light Framing*)

Structural Lumber lumber that is intended for use where allowable strength or stiffness of the piece of lumber is known

Structural Spans the spanning capability of horizontal elements determines the spacing of their vertical supports; the relationship between the span and spacing of columns, beams, slabs, and load-bearing walls influences the selection and dimension of materials for that structural system; timber planks and joists may span about 20 feet, laminated beams up to 80 feet, and wood trusses 100 feet or more; steel decking may span about 16 feet, wide-flange beams up to 60 feet, and open-web joists up to 90 feet; reinforced concrete slabs may span about 18 feet; joist slabs about 36 feet; precast planks about 40 feet, and precast tees 100 feet or more; greater beam depth allows greater structural spans, which when exposed is easily identifiable to firefighters **(Figure S-12)**

Structural Steel steel members of various shapes used as load-carrying members in a structure, and in the superstructure of high-rise buildings, because it is stronger and more durable than any other single material; it is also used in situations where the structure is required to be fire resistive/noncombustible

Structural-steel Framing girders, beams, and columns are used to construct a skeleton frame structure that transfers the gravity and lateral loads down to the foundation system; open-web joists, metal decking, diagonal bracing, moment-resisting connections, and non–load-bearing curtain walls complete a typical steel frame high rise

Structural-steel H-Pile a steel structural member whose cross section forms the letter H; it is driven into the ground to a depth determined adequate to support a building or other large structure

Structural-steel Tubing hollow lengths of structural steel in the form of round, square, or rectangular tubes used to fabricate piping-system support members and other structures where larger members, such as I-beams, are not required for the loads

Structural System a building is designed and constructed to support and transmit applied gravity and lateral loads safely to the ground without exceeding the allowable stresses in its members; the superstructure is the vertical extension of a building above the foundation: columns, beams, and load-bearing walls support floor and roof structures; the substructure is the underlying structure that forms the foundation of a building

Structural Tees precast concrete column and beam units that are connected with steel angles and plates

Structure a building or object constructed for a particular use; building materials fastened together to form an integral unit

Strut an intermediate structural member used as a brace between other members; the strut adds to the strength of the other members by providing a load path between them **(Figure S-13)**

Stucco a mixture of portland cement, lime, sand, and water used as an exterior surface covering for buildings, applied in a series of three coats to a total thickness of 3/4 inch over block, brick, or wood frame wall with metal wire or lath

Stud a wood or metal vertical framing member that makes up the walls and partitions in a frame building, usually 16 or 24 inches apart; two-by-four wood studs have a maximum spacing 16 inches on center and may go up to 14 feet in height; two-by-six wood studs have a maximum spacing 24 inches on center and may go up to 20 feet in height, except in a two-story structure, in which case studs are on 16-inch centers; studs carry vertical loads, and sheathing or diagonal bracing stiffens the plane of the wall **(Figure S-14)**

Stud-wall Sheathing includes rated panel, gypsum, fiberboard, and rigid foam plastic types, and sizes vary depending on type from 2 × 4 to 4 × 14 feet; exterior siding is secured to the exterior of the wall sheathing

Sturd-I-Floor a brand name for a type of plywood subfloor panel used as a base for interior resilient flooring

Styrofoam a trademark name for a lightweight, rigid polystyrene plastic insulating board

Subfloor the layer of flooring that is fastened directly to the floor joists to which the finish flooring is applied; it is constructed of plywood sheathing or 2 × 6 planks and may be a single or double layer **(Figure S-15)**

Subflooring the structural material that spans across floor joists and provides a base for the finish flooring, typically constructed of plywood, OSB, waferboard, particleboard, or 2-by planks; the joist and subfloor assembly can also be used as a structural diaphragm to transfer lateral forces to shear walls

Subgrade soil structure below grade level, before any surfacing material is placed; compacted soil on which a concrete slab or other structure is placed

Sump a low spot or depression below grade where water collects

Superstructure portion of a building that distributes and carries the load: vertical load-carrying systems

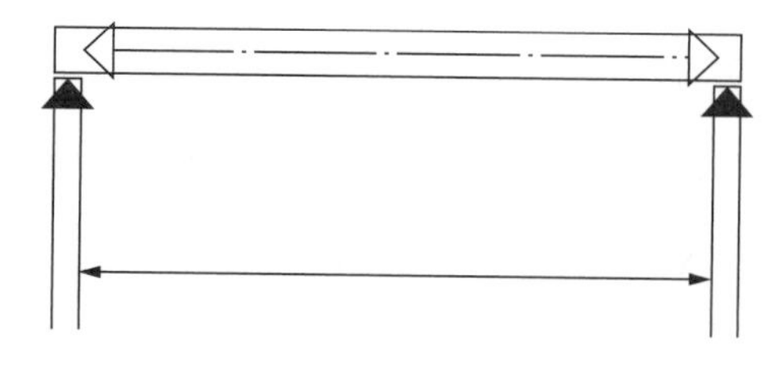

The spanning capability of horizontal elements determines the spacing of their vertical supports. This fundamental relationship between the span and spacing of structural elements influences the dimensions and scale of the spaces defined by the structural system of a building. The dimensions and proportions of structural bays, in turn, should be related to the programmatic requirements of the spaces.

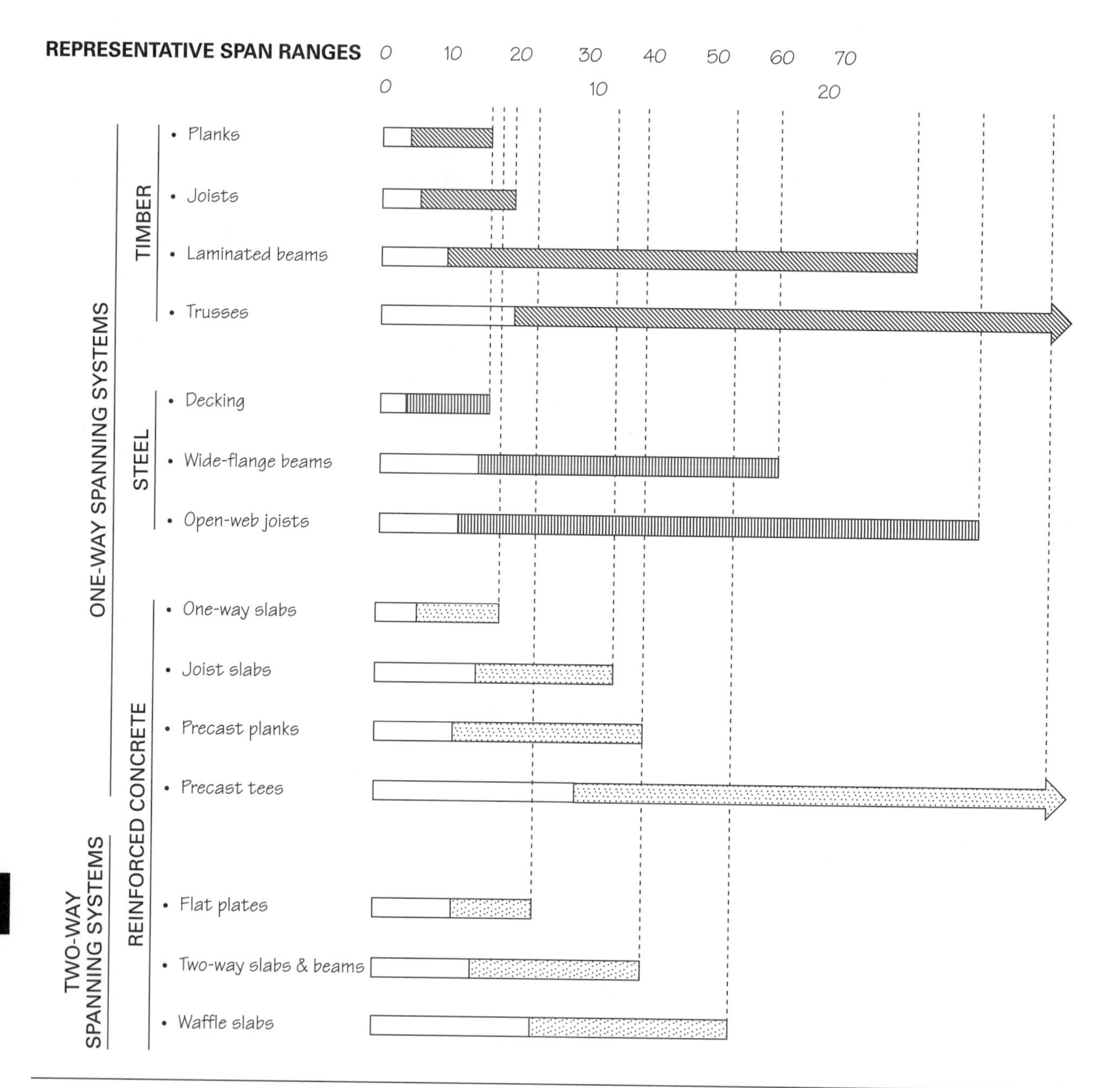

Figure S-12 Structural Spans. (*Courtesy of* Building Construction Illustrated, Third Edition, *by* Francis D.K. Ching, *© 2008, Reprinted with permission of John Wiley & Sons, Inc.*)

S

Figure S-13 Strut.

Figure S-14 Stud.

Figure S-15 Subfloor.

include bearing walls, frames, and combinations; horizontal load-carrying systems include roof and floor support systems **(Figure S-16)**

Figure S-16 Superstructure.

Surface the top layer or external exposed area of an object or structure

Surface Raceway a preformed rectangular tubing inside which electrical wiring can be routed; is usually mounted on wall or ceiling surfaces when it is impractical to run wiring inside a wall; also called *surface metal raceway*

Surface-to-Mass Ratio exposed exterior surface area of a material divided by its weight; an 8 × 8 inches timber will withstand fire impingement longer than two 2 × 4 inches members; engineered and prefabricated lightweight wood trusses and I-beams have a greater surface-to-mass ratio than a conventional, solid-sawn joist or rafter and thus will fail earlier when unprotected from fire

Surfacing adding finish material, such as stucco, to the exterior surface of a wall

Suspended Ceilings lightweight, nonstructural ceiling panels or tiles supported by a metal framework suspended by wire hangers several inches

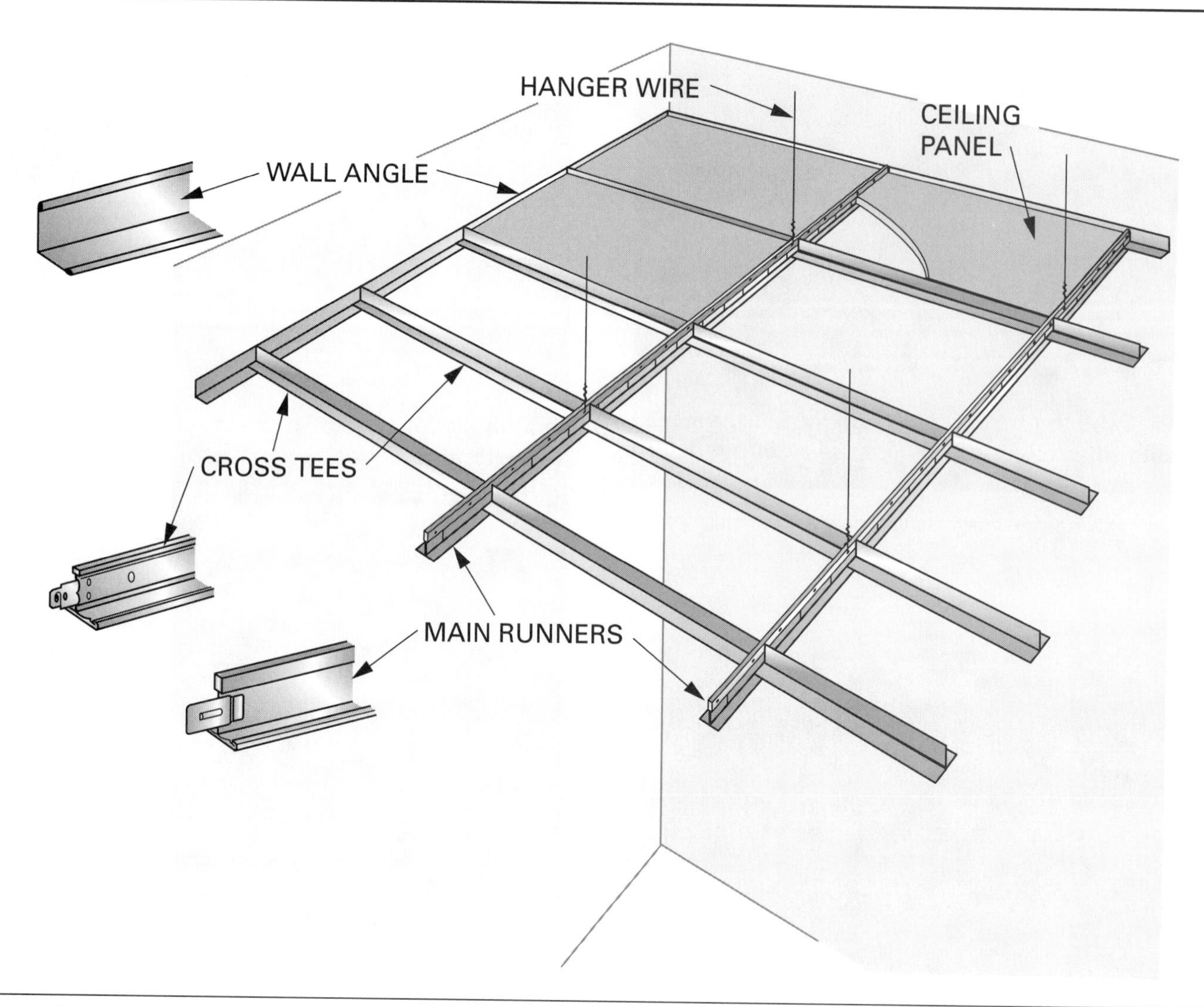

Figure S-17 Suspended Ceilings.

or feet below the supporting roof or floor beams; also known as a *hanging* or *dropped ceiling*, and the concealed space above is sometimes called a *cockloft* or *dropped ceiling*; a grid of main channels suspends tiles or runners with cross tees and splines hung by wires from the overhead floor or roof structure; some ceiling-suspension systems integrate lighting and/or air-handling systems **(Figure S-17)**

Suspended Diaphragm a structural decking system for roofs and floors designed to provide a strong surface

Suspended Load load that is held in place by attachment to something above it

Swag hanging down or drooping from one point to another, as a hanging ceiling light

Symmetrical balanced and evenly proportioned

System an assembly of parts and components that work together to perform a function, such as an electrical system or plumbing system

S

T-Beam a monolithic reinforced concrete construction in which a portion of the slab on each side of a beam acts as a flange in resisting compressive stresses, and the portion of the beam projecting below the slab serves as a web or stem in resisting bending and shear stresses

TJI a trade name for wooden I-beam

Tail Cut a cut made at the overhanging end of a rafter to trim it to the proper length and angle

Tail Joist a short joist that butts against a beam or header, or that runs from a header to a support girder

Tail Plumb Cut a vertical cut made on the end of the rafter overhang to trim the rafter to the same length as the other rafters

Tangent a line or a plane that intersects a curve or a surface at one point

Taper a gradual, sloped reduction in thickness, width, or diameter, as in a wedge or a cone

Tapered Edge the long edge of a gypsum drywall panel, which is tapered to form a shallow valley where two panels abut

Tar a dark, heavy petroleum residue remaining after other products have been distilled off, used in asphalt paving and as a sealant for objects that are in contact with high amounts of moisture or water

Tarmac a trade name for an asphaltic binder used in paving

Tees as a steel beam, has one flange and a solid stem for the web; also used to describe precast concrete units used to support floors and/or roof systems that typically span 30 to 120 feet **(Figure T-1)**

Temper the degree of hardness and elasticity of a metal, chiefly steel, as a result of blending, admixing, or application of extreme temperatures

Temporary intended for use at one location for a specified period of time; for structures, typically 180 to 360 days or less

Tendons high-strength steel cables, bars, or bundled strands used in prestressed concrete

Tenon construction that utilizes a tongue, called a *tenon,* and the opening into which it fits, called a *mortise,* to join two structural elements

Tension a force that pulls materials apart in both the vertical and horizontal plane; when stress is applied near the bottom of a beam or truss, it tends to elongate due to tension

Tensioned-membrane Structure a membrane structure incorporating a membrane and a structural support system—such as arches, columns and cables, or beams—so that the entire assembly acts together to resist loads (see also *Stress* **(Figure S-13)**)

Terra Cotta a kiln-fired tile material composed of clay and sand

Thermal Properties the behavior of a material when subjected to changes in temperature

Through-penetration Fire Stop material, device, or construction installed to resist the passage of flame, heat, and hot gases through openings that penetrate the entire fire-resistive assembly to accommodate cables, conduit, or pipes; designed and tested to resist the spread of fire through penetrations for a prescribed period of time **(Figure T-2)**

Thermoplastic Material plastic material capable of being repeatedly softened by heating and hardened by cooling; in the softened state, it can be repeatedly shaped by molding or forming

Thermoplastic Membrane Roof Covering a sheet membrane of polymers whose chemical composition allows them to be welded together by either heat or solvent

Thermoset permanently hardened by heat application

Thermoset Materials a plastic material that is substantially infusible and cannot be softened and formed after having been cured by heat or other means

Threshold a piece with chamfered edges placed on the floor under a door; also called a *sill*

Tie in concrete formwork, a device used to tie two sides of a form together; may also be a metal strip to tie a masonry wall to the wood sheathing

Figure T-1 Tees. *(Courtesy of Precast/Prestressed Institute.)*

Figure T-2 Through Penetration Fire Stop.

T

Tieback Anchors steel anchors grouted into holes drilled in the excavation wall to hold the sheeting, thus reducing the number of braces required

Tied Wall a wall braced by or supporting a roof structure to which it is fastened

Tie Rod a steel bar threaded or secured at the ends to support structural elements; often used to tie walls together or to adjust trusses to a desired level **(Figure T-3)**

Tilt-Up Construction construction in which precast concrete slabs are tilted or lifted into position by a crane onto footings or piers; concrete perimeter walls serve as bearing and shear walls; typically the precast units are prestressed, and they have been built for rapid completion, which saves time and economy (less cost in constructing and stripping vertical wall forms); full-sized panels may be up to 15 feet wide

Figure T-3 Tie Rod.

and 5 to 11 inches thick; tilt-up buildings rely on the connections between wall panels and the roof diaphragm for stability and lateral shear transfer; roof construction is often wood framing with plywood diaphragms but may also be steel bar joists with metal decks; the roof structure acts as a large diaphragm, supported vertically on interior columns; many one-story industrial buildings and two- and three-story office park structures are of tilt-up construction; many pre-1975 tilt-up buildings in California have weak ties between walls and roofs and have been required to be retrofitted with earthquake-resistant ties **(Figure T-4)**

Timber a structural lumber product 5 or more inches in the least dimension, often used for columns, beams, and stringers

Timber Trusses a heavier wood truss, assembled by layering multiple members and joining them at the panel points with split-ring connectors. Timber trusses are capable of carrying greater loads than truss rafters and are spaced further apart, usually 8 feet on center, depending upon the spanning capability of the roof decking or planking; when purlins span across the trusses, the truss spacing may be increased up to 20 feet. Wooden shaped trusses **(i.e. Bowstring Truss)** may span 40 to 150 feet, wooden flat trusses 40 to 110 feet; composite trusses have timber compression members and steel tension members

Top Chord the highest member of a truss, which is connected to other truss members

Top Plate the uppermost horizontal board nailed to the stud frame **(Figure T-5)**

Torsional Load a load parallel to the cross section of the supporting member that does not pass through the long axis; torsional loads cause structural elements and materials to twist (see also *Stress* **(Figure S-10)**)

Townhouse a single-family dwelling constructed in attached groups of three or more units in which each unit extends from the foundation to the roof and has open space on at least two sides

Transfer Beam used to support vertical loads laterally when a column cannot be aligned

Transmission of Loads loads are transmitted from where they are applied to the ground; roof and floor supporting systems direct loads to walls or frames, which direct those loads to the ground **(Figure T-6)** (see also *Load Paths*)

Tread the horizontal member of a step; tread depth is the horizontal distance from the front of a tread to the back, including any nosing

Trim door and window frames and similar decorative or protective materials used in fixed applications, such as crown molding, chair rail, baseboard, and handrail

Trimmer typically refers to a vertical wood member along the inside of a window or floor opening, parallel to the main frame members, that supports the header beam

Trunnel a wooden peg or wedge used to fasten mortise and tenon joints

Truss a framed structure consisting of a group of triangles arranged in such a manner that the loads applied at the points of intersection will cause only direct stresses (tension or compression) in its members; such a frame is based on the geometric rigidity of the triangle and is composed of linear members subjected only to axial, tension, or compressive forces; loads applied between these points cause flexural (bending) stresses; the increased depth of trusses allows them to span greater distances than steel beams and girders; components include chords, struts, ties, and panel points; some examples are pitched trusses, which have inclined top chords, and flat trusses, which have parallel top and bottom chords and raised-chord trusses that have a bottom chord raised above the level of the supports; Warren, Howe, and Pratt trusses may be flat or pitched; a *composite truss* refers to a truss having timber compression members and steel tension members; a *truss rod* is a metal tie rod serving as a tension member in a truss or trussed beam; a *trussed beam* is a timber beam stiffened by a combination of diagonal truss rods and either compression struts or suspension rods; Howe trusses have vertical web members in tension and diagonal web members in compression, Pratt

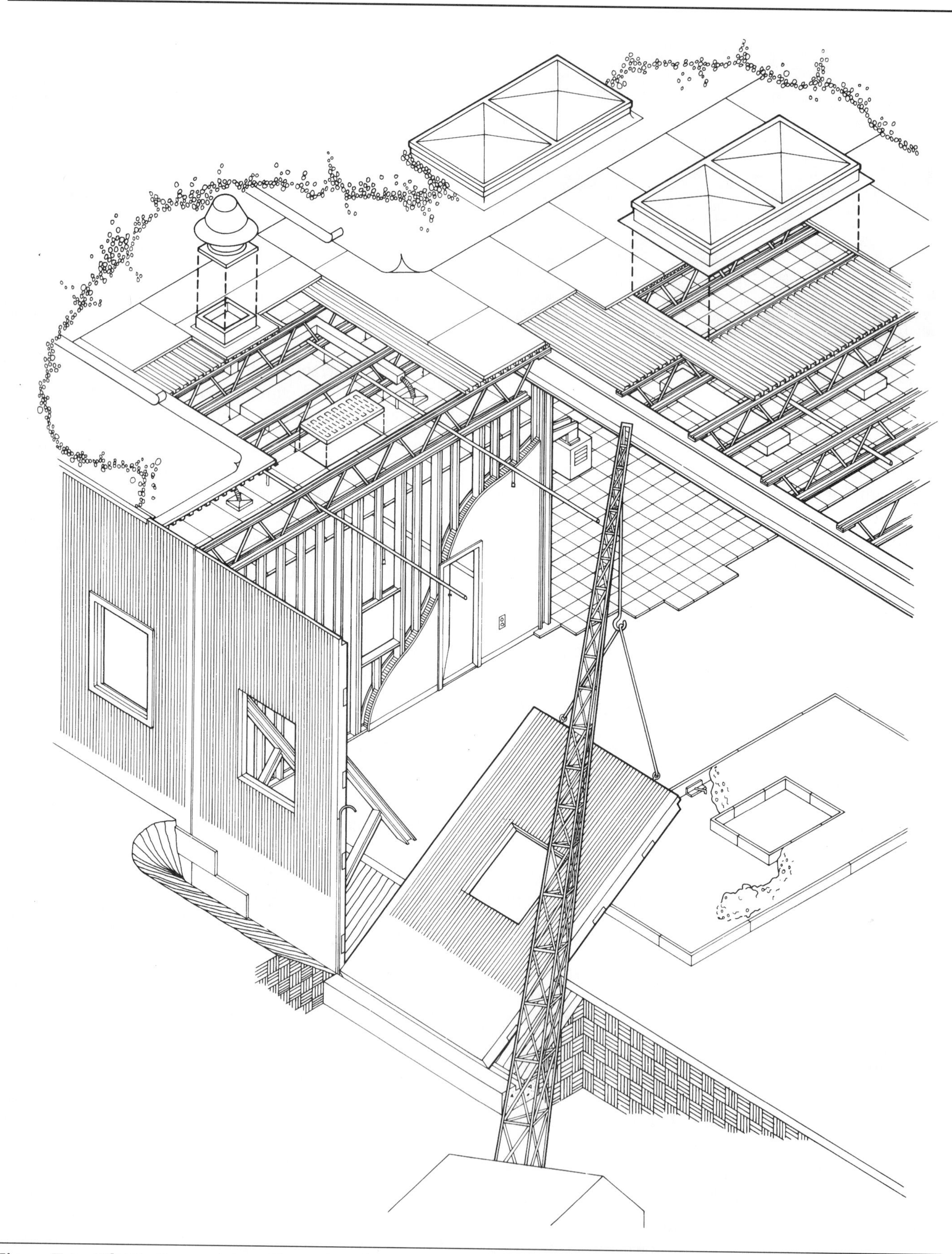

Figure T-4 Tilt-Up Construction. *(Reproduced with permission from* The Building Systems Integration Handbook, Richard Rush, *ed., Butterworth-Heimann Publishers, Newton, Mass., 1986.)*

Figure T-5 Top Plate.

Figure T-6 Transmission of Loads.

trusses have vertical web members in compression and diagonal web members in tension, and Warren trusses have inclined web members that form a series of equilateral triangles. Peaked trusses are Bowstring, Belgian, Fan, Fink, Scissors **(Figures T-7a and b)** (see also *Wood Trussed Rafters* **(Figures W-10)**)

Trussing the rigid members forming a truss, subject to axial forces proportional to the rise of the structure formed by trusses; although rigid in its own plane, a truss must be braced in a perpendicular direction to prevent lateral buckling

Truss Frame a wood-frame engineered construction in which roof and floor trusses and studs are tied in a unitized frame; the small dimensional lumber used (2 × 4 or smaller) burns faster than larger solid lumber

⚠ **Truss Loft** the void between the top-floor ceiling and roof, where hidden fire can spread; also known as a *truss void* **(Figure T-8)**

Truss Plate a steel plate used to strengthen the joints in truss assemblies

Trussed Beams timber beams stiffened by a combination of diagonal truss rods and either compression struts or suspension rods

Trussed Rafters preengineered and shop-fabricated monoplanar trusses, typically of 2 × 4 or 2 × 6 lumber, that span 20 to 32 feet; typical spacing is 2 feet on center but may vary up to 4 feet; truss members are connected with metal-toothed plate connectors and may be supported by a stud-framed or masonry bearing wall, timbers, or steel beams; roof sheathing is similar to conventional rafter framing, and a ceiling may be applied directly to the bottom chords **(Figure T-9)** (see also *Wood Trussed Rafters* **(Figure W-8)**)

Truss Rods metal tie rods that serve as tension members in a truss or trussed beam

Tubes cylindrical, square, or rectangular steel shapes most often used as columns

Turnbuckle piece of hardware that allows two threaded rods to be tightened or loosened by rotating a threaded sleeve; because turnbuckles and threaded rods are made of steel and are typically used to keep structural elements aligned, firefighters need to report these features during fire-ground activities

Two-way Slab and Beam two-way slab and beam concrete floor construction is of uniform thickness and may be reinforced in two directions, cast integrally with supporting beams and columns on all four sides of square or nearly square bays; this construction is effective for medium spans (15 to 40 feet) and heavy loads or when a high resistance to lateral forces is required; *two-way waffle slabs* are reinforced in two directions, which allows the floor to carry heavier loads and span longer distances than flat slabs; suitable for spans of 24 to 54 feet, longer spans may be possible with posttensioning; *two-way flat plate slabs* are reinforced in two or more directions and supported directly by columns without beams or girders; used for apartment and hotel construction due to lower floor-to-floor heights. Suitable for light to moderate loads over relatively short spans (12 to 24 feet); *two-way flat slab* is a flat plate thickened at its column supports to increase its shear strength; a *drop panel* is the portion of a flat slab thickened around a column head to increase its resistance to punching shear; it is suitable for relatively heavy loads and spans from 20 to 40 feet; (see also *Waffle Slab, Flat Slab Concrete Frame, Flat Slab Floor,* and *Concrete Floor System* **(Figure C-15)**)

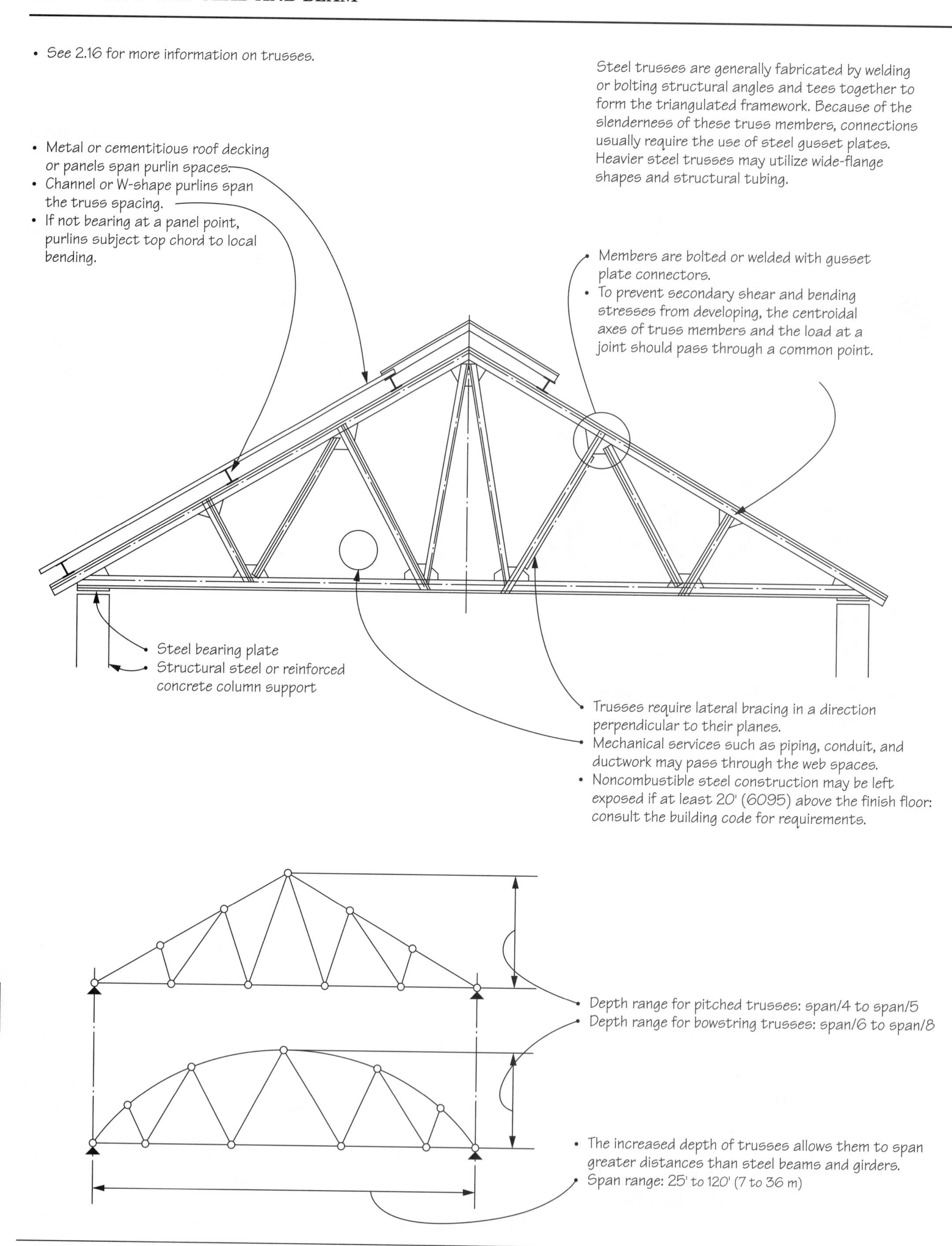

Figure T-7a Truss. *(Courtesy of* Building Construction Illustrated, *Third Edition, by* Francis D.K. Ching, *© 2008, Reprinted with permission of John Wiley & Sons, Inc.)*

Figure T-7b Truss. *(Courtesy of Kathleen Siegel.)*

Figure T-8 Truss Loft. (*Courtesy of Kathleen Siegel.*)

Figure T-9 Trussed Rafters.

Types of Construction building codes classify the construction of a building according to the fire resistance ratings in hours of its major components, which includes the structural frame; exterior bearing and nonbearing walls; interior bearing walls and permanent partitions; floor, ceiling, and roof assemblies; and stairways and shaft enclosures; five classifications are used in the major building codes, but they may differ in details of walls, structural frames, and floors

Type I Construction also known as *fire-resistive construction,* Type I buildings are of noncombustible materials and have a structural framework of protected steel, reinforced concrete, or reinforced masonry; bearing walls, columns, and beams are required to have 3- to 4-hour fire resistance, floor construction is required to have a fire resistance of either 2 or 3 hours, and roof construction must have either 1 or 2 hours of fire resistance but may be unprotected if they are a certain height above the floor **(Figure T-10)**

Type II Construction also known as *noncombustible construction,* Type II buildings are similar to Type I buildings except for a reduction in the required fire-resistance rating of building elements. Structural members such as walls, columns, beams, girders, trusses, arches, floors, and roofs are of approved noncombustible or limited combustible materials and may be either protected or unprotected construction; many Type II buildings with roofs 20 feet or more in height may not be required to provide fire protection for the roof framing and decking; fire-retardant–treated wood members are permitted under some conditions, as are some combustible materials; combustible materials are not permitted within the concealed spaces in Type I and II buildings unless they meet a Class A rating according to the NFPA 5000; combustible mechanical, plumbing, and electrical components must be installed in accordance with the applicable code **(Figure T-11)**

Type III Construction also known as *ordinary construction,* Type III buildings have noncombustible exterior walls and a structural framework of steel, concrete, masonry, or wood **(Figure T-12)**

Type IV Construction also known as *heavy timber construction,* Type IV buildings have noncombustible exterior walls, an interior structural framework of solid or glue-laminated timber of specified minimum sizes, and thick floors planks with no concealed spaces; allowable nominal dimensions for columns that support floor loads cannot be less than 8 inches; if they support only roof loads, they cannot be less than 6 inches in width and 8 inches in depth; wood beams and girders supporting floor loads cannot be less than 6 inches in width or less than 8 inches in depth **(Figure T-13)**

Type V Construction also known as *wood frame construction,* Type V buildings have building elements with the fire-resistance ratings specified in the building code; this type of construction features exterior walls, bearing walls, floors, columns, beams, girders, trusses, arches, and roofs that are made completely or partially of wood or other approved materials of a smaller dimension than those used for Type IV construction **(Figure T-14)**

TYPE I BUILDINGS *("Fire Resistive")*	
DEFINITION	Structural components shall be of noncombustible materials such as steel, iron, concrete or masonry. Framing resist fire for three to four hours; floors, roofs and partitions are designed to meet lesser periods of fire resistance.
DESCRIPTION	Typically protected steel frame and/or reinforced concrete. Two inch coating of spray-on gypsum/concrete/mixes on steel frame.
DESIGN	Typically mid to high rises since the turn of the century. Contemporary high rises require fire protection systems (detection, alarm and sprinklers/standpipes). Various occupancies include business, residential hospitals, retail and commercial.
STRUCTURAL ELEMENTS	Steel supporting skeleton with I, H and tube columns and beams with spandrel girders to tie together the exterior wall columns. They carry the weight of the panel or curtain walls and resist wind and lateral forces. Type I buildings may also have load-bearing masonry walls (up to 20 stories) and prestressed concrete members. Floors and roofs typically steel decks with lightweight concrete topping or precast concrete panels.
FIRE SPREAD CHARACTERISTICS	Building does not contribute to fire nor is it easily damaged, even by extended periods of exposure. Interior fire spread primarily due to burning contents. Open floor area design and HVAC systems may allow smoke and heat to travel throughout the space. Floor to floor lapping through windows and panel walls. Excessive heat conditions have been noted in Type I building fires.
COLLAPSE HAZARDS	Rarely collapse but may be gutted. Sagging ceilings, ductwork, utilities likely. Exterior glass/metal panels may fail and present falling dangers and vertical fire spread to adjacent floors.
FIRE GROUND CONSIDERATIONS	Support automatic sprinkler and standpipe systems. Consider multiple high-flow hand lines or portable monitors. Consider heat and smoke exposures to adjacent structures. Concrete floors/ceilings may spall due to high heat conditions exposing steel members. Building height and area may be very large and require high fire flow requirements. Time factor required to gain access to fire floor(s) may necessitate large numbers of firefighters for initial fire attack.

Figure T-10 Type I Construction.

TYPE II BUILDING *("Noncombustible")*	
DEFINITION	Structural components are of noncombustible materials (steel, concrete, masonry) with limited use of combustible materials. Similar to Type I with a reduction in the required fire-resistance rating for structural elements.
DESCRIPTION	Typically reinforced concrete or masonry walls with protected or unprotected steel beams to support the roof. May or may not have automatic fire sprinklers. Typical examples include unprotected steel columns and open web bar joists that support steel or wood roof decks.
DESIGN	May be low, mid or high rise. Steel sided, single story unprotected steel framed shops, warehouses, industrial and commercial occupancies are common. Open interior areas with few windows and doors.
STRUCTURAL ELEMENTS	Structural members contribute little or no fuel to fires. Typical examples include steel frames with metal exterior walls, steel frames with concrete block walls and concrete bearing walls supporting metal roof systems. Steel bar joists can be used for spans up to 60'.
FIRE SPREAD CHARACTERISTICS	Interior fire spread due to primarily burning contents. Large open areas with or without ceilings and/or draft stops allow for heat and smoke travel. Metal wall panels have less ability to resist fire and heat than masonry units or gypsum wallboard. Insulated metal roof deck fires need to be cooled from underside.
COLLAPSE HAZARDS	Unprotected Type II buildings will not provide structural stability under fire conditions. Failure of unprotected steel from the heat of burning contents should be anticipated. Interior height of roof supporting system, intensity and duration of fire exposure will dictate time to failure. Extent of roof collapse dependent upon extent of fire, exposure and damage.
FIRE GROUND CONSIDERATIONS	Support automatic sprinkler and standpipe systems. Heat needs to be released and/or the steel cooled to reduce thermal conductivity. Steel bar joist may fail in 5–10 minutes, which can cause roof deck failure. Cool interior steel members and decking. Consider high heat conditions, warping/buckling/failure of structural members if timely and adequate cooling streams are not available. Consider multiple high-flow hand lines or portable monitors. Consider heat and smoke exposure to adjacent structures. Consider defensive operations for low to mid-rise unprotected structures.

Figure T-11 Type II Construction.

TYPE III BUILDING *("Ordinary Construction")*	
DEFINITION	Exterior walls/structural members are of noncombustible or limited combustible materials. Interior structural members are completely or partially constructed of wood.
DESCRIPTION	Typically reinforced concrete or masonry exterior load bearing walls with a variety of different materials and techniques for interior members. Older Type III buildings may be identified by deep windows and doors and thick brick walls.
DESIGN	Typically low to mid rises often with narrow widths and deeper lengths. Older Type III buildings have masonry walls with wood joisted floors 12 to 16 inch centers with spans up to 25 feet. Newer Type III buildings include tilt-ups with light-weight steel bar joists. Occupancies include offices, hotels, commercial/industrial buildings, strip malls and storage facilities.
STRUCTURAL ELEMENTS	Variable. Includes URM walls with wood joists and timber trusses; concrete tilt-up walls with open web tube trusses and/or steel bar joists; wood floors and roofs supported by interior wood or steel columns. Older Type IIIs have floors and roofs that may have poor connections and that be in reinforced with steel tie rods and spreader plates.
FIRE SPREAD CHARACTERISTICS	Interior structural members may be combustible and contribute to fire spread. Concealed spaces and interconnected voids between floor and ceilings joists and interior walls allow for fire spread, particularly with multiple dropped ceilings. Fire extension is possible in party walls with joist recesses and butt joints.
COLLAPSE HAZARDS	Structural stability cannot be assumed when the interior structural components are combustible. Large trusses may fail after prolonged fire exposure and push walls out. When interior wood columns, beams or floors are lost, the exterior wall loses stability and may collapse. Unprotected steel columns and bar joists will perform poorly under fire conditions and may fail early. Parapet walls require immediate collapse zone designation.
FIRE GROUND CONSIDERATIONS	Support automatic sprinkler and standpipe systems if available. Older Type III buildings may be one of the most hazardous structures during fire attack. Unreinforced masonry walls with lime/sand motar, fire cut joists, timber truss roof systems, common/undivided attics and parapet walls all present challenges to fire fighting forces. Consider collapse zones and ensure proper apparatus spotting.

Figure T-12 Type III Construction.

TYPE IV BUILDING ("Heavy Timber")	
DEFINITION	Exterior walls and structural members that are non-combustible or limited combustible materials. Interior structural members are completely or partially constructed of large dimensioned wood (solid or laminated) with no concealed spaces.
DESCRIPTION	Typically reinforced concrete or masonry exterior load bearing walls with columns, beams and girders not less than 8 inches in any dimension. Rarely built as new construction, these buildings were popular in the 19th century for manufacturing, factories, mills, warehouses, churches and schools.
DESIGN	Typically low to mid rises with no concealed spaces. Large, deep window and door spaces indicate thick walls used to carry loads. Recent conversions to business, residential and assembly occupancies.
STRUCTURAL ELEMENTS	Exterior walls typically masonry often will wall columns. Beams and girders supporting floor loads typically 6 × 10 minimum; beams and girders supporting roof framing and loads typically 4 × 6 minimum. Tongue and groove planks for floors and roof decking typically 3 inch or larger. Eight inch laminated timbers common for beams and arches as well as heavy timber trusses.
FIRE SPREAD CHARACTERISTICS	Large interior structural members are combustible and contribute to fire spread. Large timbers are slower to ignite but present massive amounts of fuel to burn. Heavy fire loading in large, open floor areas provide vertical and horizontal fire spread.
COLLAPSE HAZARDS	Structural stability cannot be assumed when the interior structural components are combustible. Large trusses may fail after prolonged fire exposure pushing heavy masonry walls out. When interior wood columns, beams or floor are lost, the exterior wall loses stability and may fail in later stages. Parapet wall failure; establish collapse zones early on.
FIRE GROUND CONSIDERATIONS	Support automatic sprinkler and standpipe systems if available. Extreme radiated heat and exposure issues. Portable and aerial master steams with adequate water supply necessary to extinguish deep-seated fires. Renovations and conversions to other uses may create concealed spaces similar to Type III and V buildings. Consider collapse zones and ensure safe operating positioning for personnel and apparatus.

Figure T-13 Type IV Construction.

T

TYPE V BUILDING (*"Combustible wood frame"*)	
DEFINITION	Exterior and interior walls, floors, roofs and supports made completely or partially of wood or other approved materials of smaller dimensions than those used for Type IV construction. Also called *light frame construction*.
DESCRIPTION	The most common type of construction due to low cost, ease of construction and availability. Most codes limit wood frame buildings to not more than three floors.
DESIGN	Typically low rise buildings. Styles include Ranch, Victorian, Craftsman, Eichler and many others.
STRUCTURAL ELEMENTS	Structural elements may be grouped as solid wood or engineered/manufactured lumber. Wood studs typically 2 × 4 or 2 × 6 on 12, 16 or 24 inch centers; solid joists 2 × 4 up to 2×16 inches on 16 or 24 inch centers; solid rafters 2 × 6 up to 2 × 16 inches on 16 or 24 inch centers. Engineered wood products such as wooden I-beams, parallel chord trusses, plywood and OSB are common in new construction.
FIRE SPREAD CHARACTERISTICS	Vulnerable to fire and failure due to combustibility. Fire will spread vertically and horizontally. Enclosed walls, floor/ceiling/roof areas create concealed spaces and voids.
COLLAPSE HAZARDS	Structural members are combustible and thus will fail when exposed to fire. Remodeling, alterations and renovations by unqualified workers may bear components that could cause early and widespread collapse. Lightweight construction materials and members will have earlier failure than full-dimensional members. Roofs are not designed to carry the same loads as floors. Roofs typically are subject to earlier failure than floors or walls.
FIRE GROUND CONSIDERATIONS	Construction type, style, ages and materials may dictate tactics. Fire damage to wooden structural members = loss of strength = change in forces and application of design loads = structural failure and collapse. Support automatic sprinkler and standpipe systems if available.

Figure T-14 Type V Construction.

UL Underwriters Laboratories publishes safety standards for a variety of industries; UL 263 is "Fire Tests of Building Construction and Materials"

Underlayment in flooring, a base material that provides a smooth, level surface for the direct application of nonstructural resilient flooring; underlayment may be a separate layer of plywood or fiberboard, or it may be combined in a single thickness with the subfloor panel; in roofing, underlayment is typically 15- or 30-pound asphalt-saturated felt that protects the roof sheathing from moisture until the shingles are applied

⚠ **Undesigned Load** a load that the building designer did not anticipate or plan for (see also *Load Paths* **(Figure L-7)**)

⚠ **Undivided** refers to space typically in attic, cockloft, and mansard areas that allow fire to travel throughout unobstructed

Uniform Building Code (UBC) published by the International Council of Building Officials, a model code that determines the minimum standards those buildings must meet in the interest of community safety and health

Uniform Fire Code (UFC) published by the International Council of Building Officials

⚠ **Unprotected** without fire resistance **(Figure U-1)**

Figure U-1 Unprotected.

⚠ **Unreinforced Masonry (URM)** brick, block, or stone units that are not reinforced with steel, a type of construction that performs poorly in earthquakes due to a lack of tension capacity and ductility; retrofitting of unreinforced masonry buildings with vertical and horizontal reinforcement has proved beneficial in recent seismic events; shear-wall connections at roofs and floors, threaded rods with steel bearing plates at the ends of walls, and reinforcing walls from outside the existing structure are some retrofitting methods; many URMs are of Type III construction with wood interior framing, but some larger URMs utilize timber and steel trusses for the roof support system; fire can cause truss failure, which can cause the secondary, unreinforced masonry wall collapse; many URMs also have unprotected steel columns, beams, and lintels that also lead to URM wall failure when heated by fire **(Figure U-2)**

⚠ **Unrestrained Beam End** a beam end resting on a support held in place only by gravity; examples include a fire-cut beam or a beam resting on a corbel ledge or on a girder **(Figure U-3)**

Figure U-2 Unreinforced Masonry (URM).

Figure U-3 Unrestrained Beam End.

Vacant Buildings abandoned structures, which may have disabled fire protection systems and removed structural members; these pose additional risks to firefighters, as transients may illegally occupy vacant buildings and those under demolition, and serious fires in such structures have resulted in many injuries and line-of-duty deaths

Valley an intersection of two inclined roof surfaces toward which rainwater flows

Valleys the intersections of roof surfaces; *open valleys* are exposed metal flashings, and *closed valleys* are covered with singles; for both, mineral-surfaced roll roofing is applied below the metal valley

Valley Rafter the rafter placed at the intersection of two roof slopes in interior corners; valley jack rafters and valley cripple-jack rafters describe specific rafter types based on their location in the valley

Vapor Barrier a watertight material used to prevent the passage of moisture or water vapor into and through walls

Vault an arched structure of stone, brick, or reinforced concrete forming a ceiling or roof over a hall, room, or other enclosed space

Veneer a facing attached to a wall for the purpose of providing ornamentation, protection, or insulation but not counted as adding strength to the wall; materials include masonry, plaster, plywood, and Styrofoam

Veneered Wall a wythe of masonry attached to the bearing wall but not carrying any load but its own weight; often used for a desired aesthetic appearance

Ventilation the exchange of air in a building for roofs, attics, crawl spaces, bathrooms, and kitchens; firefighters may perform vertical or horizontal ventilation to provide a more rapid removal of the products of combustion during a fire

Vinyl Siding a shaped material, made principally from rigid polyvinyl chloride (PVC), that is used as an exterior wall covering

Void an empty space between members or elements of a structure that remain unseen without removing coverings, also known as *concealed spaces*; inherent in Type III and Type V buildings, interconnected void spaces are created by trusses and steel bar joists; combustible voids are created with wood joists, floors, rafters, and roof decks **(Figure V-1)** (see also *Concealed Spaces*)

Figure V-1 Void.

Voussoirs individual, wedge-shaped pieces of an arch, the central or highest of which may be called a *keystone* (see also *Keystone*)

Waferboard a nonveneered wood panel product composed of large, thin wood flakes bonded under heat and pressure with a waterproof adhesive, typically used for sheathing and paneling

Waffle Slab a concrete slab that has ribs running in two directions forming a wafflelike grid

Wainscot a facing of wood paneling that usually covers the lower portion of an interior wall, as a wall finish for a section of wall that is different from the top section; it is often paneled or tiled, with a finishing molding that covers the seam

Waler reinforcing structural members used to brace concrete forms or piles, usually made of 2 × 3 inches or larger lumber and installed horizontally on the outsides of the forms, spanning between studs, and tied through the forms using wall or form ties

Wall one of the sides of a structure surrounding an enclosure

Wallboard a panel designed as a wall covering, typically made of a variety of materials, such as gypsum drywall and wood products

Walls any of various upright constructions presenting a continuous surface and serving to enclose, divide, or protect spaces; they transmit to the ground the compressive force applied along the top, or received at any point on the wall, and may be structural frames, concrete and masonry bearing walls, or metal and wood stud walls **(Figure W-1)**

Assorted wall terms include:

- ***Load-bearing wall*** carries a load of some part of the structure—floors, ceilings, and roofs—in addition to the weight of the wall itself; also known as *bearing wall,* a collapse of this type of wall is more serious than the collapse of a column, floor, roof beam, or nonbearing wall
- ***Non–load-bearing*** supports only its own weight
- ***Cantilever wall*** walls attached only at one end
- ***Cross wall*** any wall at right angles to any other wall
- ***Panel or curtain wall*** non–load-bearing enclosing walls on a frame building
- ***Parapet wall*** a short wall above the roofline
- ***Veneer wall*** a finished brick or stone facing used on the outside of a building to improve the exterior appearance
- ***Party wall*** a load-bearing wall common to two structures
- ***Partition wall*** an interior wall that subdivides areas of a floor; may be a bearing partition that carries a structural load or a nonbearing partition wall that supports no load other that its own weight
- ***Pony wall*** a dwarf wall for supporting floor joists
- ***Shear wall*** wall in a framed building designed to assist in resisting the forces of wind
- ***Fire Wall*** wall that separates two connected structures, or that divides portions of a large structure, to prevent the spread of fire from one structure or portion to the next

Wall Beam a structural member serving both as a wall and as a beam

Wall Bearing Structure a building in which the walls support the roof and floors and are not dependent upon a skeleton frame to support themselves **(Figure W-2)**

Wall Bracing examples include buttresses, pilasters, wall columns, and cavity or hollow walls that provide greater load-carrying capacity

Wall Column typically a concrete or masonry column designed to stiffen the wall and carry the weight of a concentrated load; provides visual clues for firefighters as to locations of beams and/or trusses

⚠ **Wall Failure** examples of line-of-duty deaths due to wall failure include:

- **Chicago IL** Nine firefighters died in a six-story commercial brick structure
- **Boston, MA** Five firefighters died in a vacant (Type III) toy factory
- **New York, NY 1956** Six firefighters killed by a collapsing parapet wall of a single story furniture showroom

Wall failure is often due to roof or floor failure **(Figure W-3)**

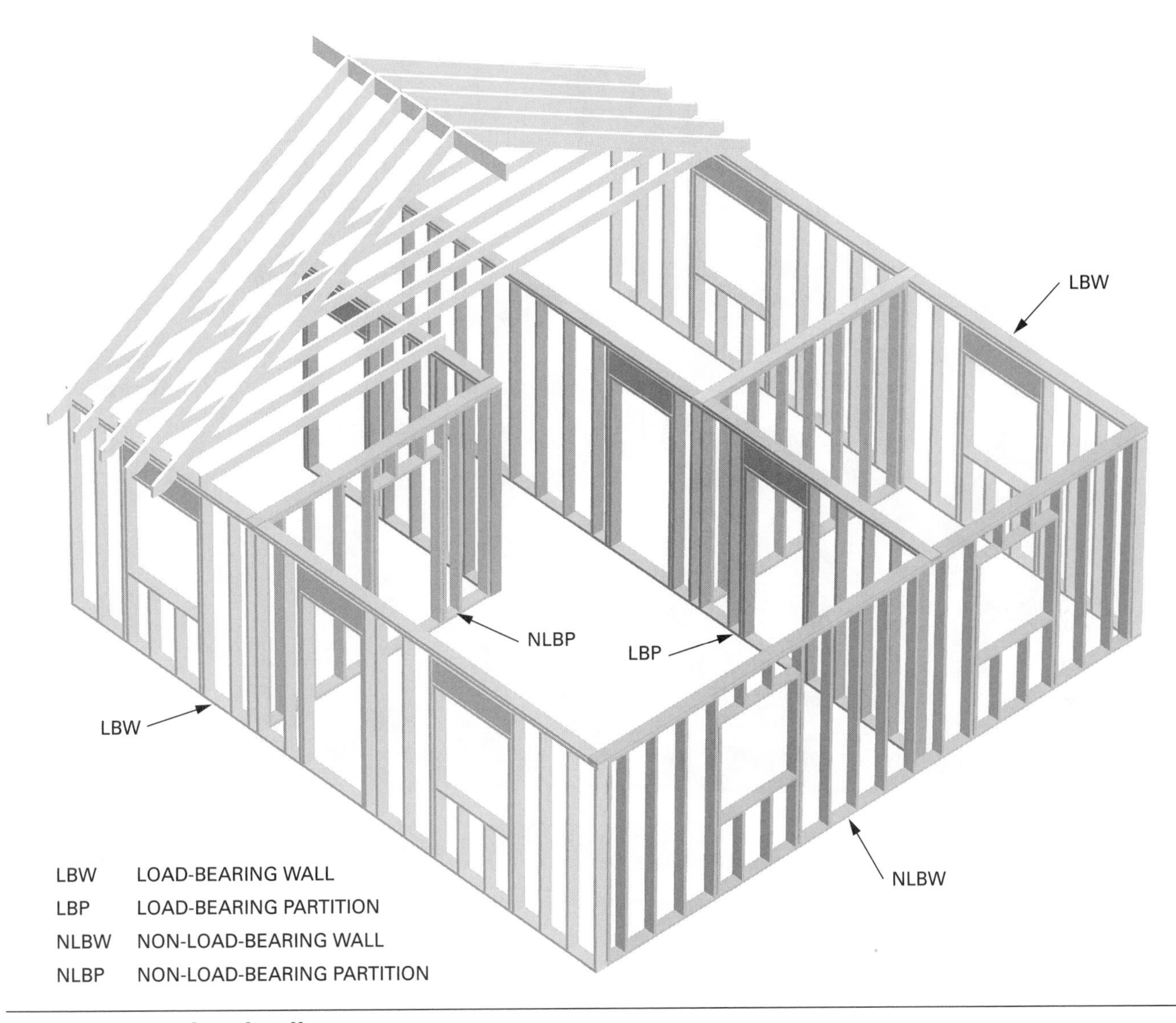

Figure W-1 Wood Stud Walls.

Figure W-2 Wall Bearing Structure.

Wall Footing a widened section at the base of a wall foundation that bears against the soil and distributes structural loads directly to the soil

Wall Panel nonbearing wall built between columns and piers wholly supported at each story; typically masonry, metal, glass, or other material is used in conjunction with structural steel or concrete frames (see also *Curtain Wall* **(Figure C-22)** and *Panel Wall* **(Figure P-2))**

Wall Plate a horizontal structural member anchored to a masonry wall; other structural members may be supported from the wall plate

Wall Sheathing exterior wall covering that may consist of boards, rated panels, gypsum board, fiberboard, or rigid foam board

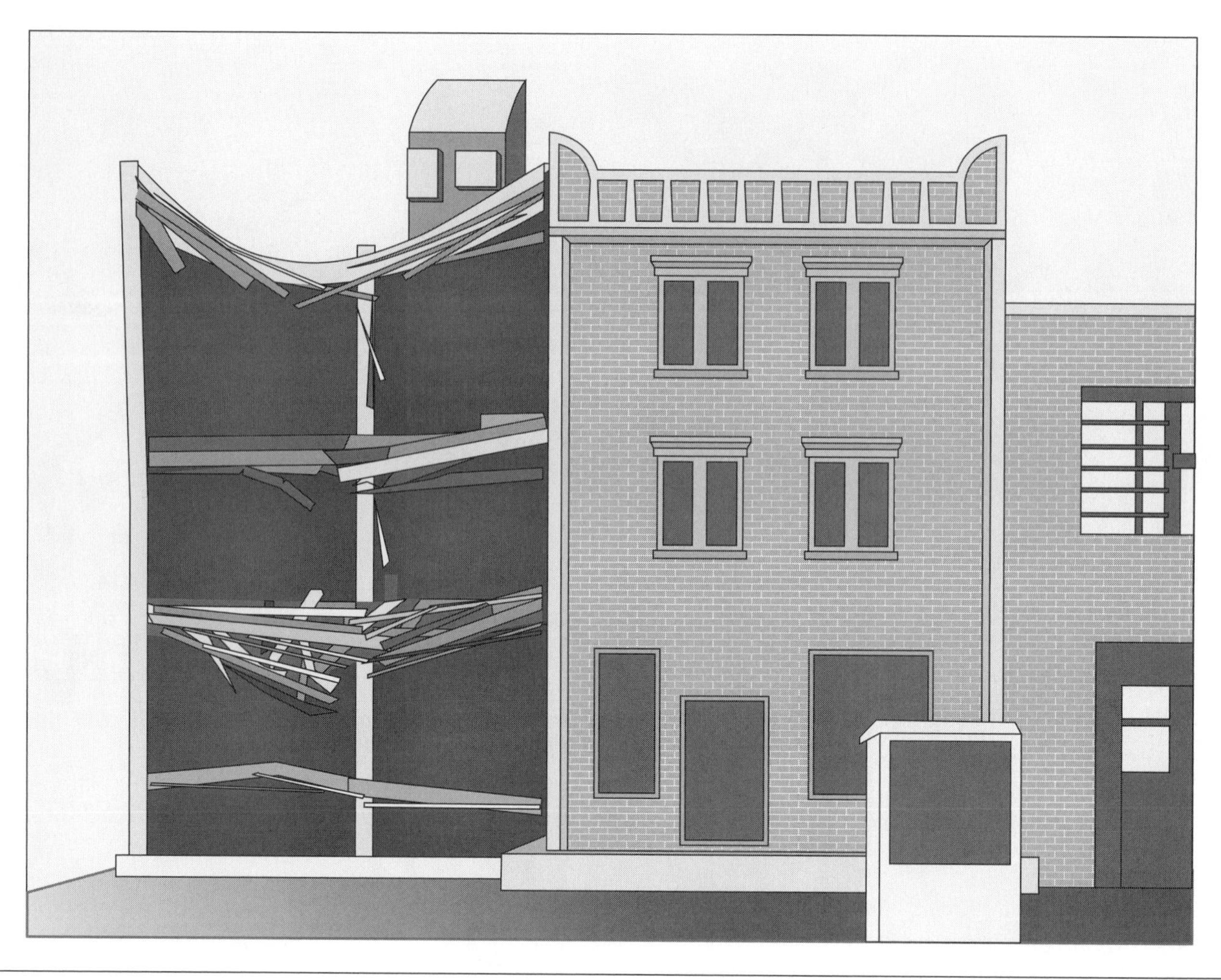

Figure W-3 Wall Failure Due to Roof/Floor Failure.

Wall Systems a building's primary vertical elements; may be homogeneous or composite construction or may be composed of linear bearing elements (posts and columns) with nonstructural panels in between; typical walls systems support either floor or roof systems above and are in turn supported by wall, floor, or foundation systems below; general groupings typically include structural frames—concrete, steel, timber, and steel/concrete frames; concrete and masonry bearing walls, which use mass for load-carrying capability that requires reinforcement to handle tensile stresses; and metal and wood stud walls, typically 16 or 24 inches on center with the studs carrying vertical loads while sheathing or diagonal bracing stiffens the plane of the wall

Wall Tie a metal piece used to tie masonry wythes to each other or to another structure; also known as a *tie*

Warp a variation from straightness in a piece of lumber

Water Hammer a noise created by a water surge, or a rapid flow change of water, in a stream system

Water Main a large diameter piping system used to transport water from a water district treatment plant to the area of use; individual water lines run from the main to homes and buildings

Water-resistive Barrier a material behind an exterior wall covering that is intended to resist water that has penetrated behind the exterior covering to prevent further penetration into the interior wall assembly

Water-vapor Retarder material or construction that adequately impedes the transmission of water vapor

Waterproof impervious to penetration by water

Waterproofed Portland Cement Portland cement with waterproof materials used to coat the cement

W–Beam a steel structural I-beam with wide flanges (the *W* stands for *wide*); also called a *wide-flange beam*

Weatherhead a rainproof conduit fitting for the service entrance head where overhead electrical power is brought into a building

Web the wide, vertical stem or central portion of an I-beam between the flanges, or in the case of a truss, the connection between the upper and lower chords **(Figure W-4)**

Figure W-4 Web. *(Courtesy of Pat McAuliff.)*

Web Connection a connection, usually made of steel, to the web of a beam **(Figure W-5)**

Figure W-5 Web Connection.

Web Member secondary members of a truss contained between chords **(Figure W-6)**

Web Stiffiner a wood block used to reinforce the web on an I-joist, typically at the ends as needed

Weep Hole a hole or holes in masonry walls that permit the passage of water to keep it from building up behind the wall and possibly undermining the foundation

Figure W-6 Web Member.

Weld to fuse metals together by melting the pieces at the interface where they are to be joined; pressure and/or a filler material may also be used to aid in the fusion

Western Framing a contemporary style of wood frame construction in which each story is built on a platform, providing fire stopping at each level (see also *Platform Framing*)

Western-union Splice a splice for joining two electrical wires together by overlapping the bare ends and twisting each around the other

Wet-pipe Sprinkler System a fire-protection system in which the water supply remains in the pipes under pressure, ready to be released through the sprinkler heads when heat is sufficient to open them

Wet Wall a masonry wall that is built using mortar

Wide-Flange Shapes I-beams that have flanges wider than standard I-beams

Wildland/Urban Interface structures located in a wildland/urban interface or wildland/urban intermix; see NFPA 1144, "Standard for the Protection of Life and Property from Wildfire"

Wind Loads force applied to a building by the wind

Wing Wall a wall extending at a right angle to another

Wired Glass two panels of glass with wire mesh embedded between them during manufacturing; the wire serves as reinforcement and prevents the glass from shattering when broken; wired glass is used primarily for security purposes, when lights are set into doors, and it may be required to prevent fire extension

Wood one of the oldest and most commonly used building materials, used for structural members such as beams, columns, girders, decking, arches, panels, trusses, and foundation pilings; temporary timber shoring and formwork are used in the construction of steel, masonry, and concrete structures; wood is also used as a finishing material and for nonstructural applications, such as for furnishing and cabinetwork; advantages of wood are good strength-to-weight ratio, excellent insulating properties, good fire resistance (in large dimension cross sections), and aesthetic appeal; wood is also a renewable resource, and wood products are biodegradable; problems related to wood use include decay, combustibility, and rapid flame spread across its surface; moisture, fungi, and insects can cause wood destruction, and roof membrane deterioration has caused structural wood members to deteriorate, leading to weakness and failure; as wood ages, it loses moisture, which causes it to shrink and weaken at its connectors; as wood dries out, its ignition temperature is lowered; when wood burns, it loses structural strength and integrity; most of the low-rise structures built today—including school buildings, residences, and commercial and apartment buildings—are of wood construction; new wood products have been developed that include glued-laminated members and panel products such as plywood, particle board, and oriented-strand board, and these improve the material's efficiency, durability, fire resistance, and structural performance; proprietary structural timber products and "preengineered" timber and wood systems are becoming commonplace; the structural integrity of wood frame construction is highly dependent on the quality of the connections (see *Connectors*); wood is also a durable structural material if proper construction procedures and appropriate maintenance are provided

Wood Beams include solid beams, glue-laminated timber beams, box beams, flitch beams, spaced beams, and built-up beams

Wood-Block Flooring flooring laid in preformed wood tiles, often with tongue-and-groove edges

Wood Columns wood columns and posts may be solid, built-up, or spaced and are loaded axially in compression; spacing is related to the desired size and proportion of the bays, and the spanning capability of the beams, joists, and decking

Wood Construction Types typical framing styles include balloon, platform, post and frame, post and beam, truss frame, plank and beam, log cabin, and A-frame; designs include ranch, bungalow, Craftsman, Victorian, Eichler, and many others (see also *Wood Frame Construction*)

Wood Decking plywood, 1-inch subfloor, or 2-inch tongue-and-groove lumber or planks that rest on the floor support system and are covered by finish flooring

Wood Floor Systems includes wood joist systems and wood plank and beam systems

Wood Frame Construction a type of building construction in which all the loadbearing structural members—the studs, plates, joists, and rafters—are made of wood; may also be referred to as *combustible frame construction* (see also *Type V Construction* **(Figure T-14)**)

Wood Joists used in light frame construction to create floor and ceiling assemblies, conventional wood joists are spaced 12, 16, or 24 inches on center, depending on the loads and spanning capacity of the wood-panel sheathing or subflooring; ceilings may be applied directly to the joists, or they may be suspended to lower the ceiling; wood stud framing, wood or steel beams, or a bearing wall of concrete or masonry may support joists; typical span ranges for solid-sawn 2 × 6 wood joists are up to 10 feet; for 2 × 8 wood joists, the span is 8 to 12 feet; for 2 × 10 joists, 10 to 14 feet; and 2 × 12 joists span 12 to 18 feet; newer, wooden I-beams are used for wooden joists and provide greater strength and uniformity but less fire resistance if unprotected (see also *Joist* **(Figure J-1)**)

Wood Lath narrow, rough strips of wood nailed to studs and covered with plaster; predominately used before gypsum board in older wood-framed structures

Wood Panel Products manufactured wood product used by the construction industry that include plywood, high-density overlay (HDO), medium density overlay (MDO), particleboard, orientated strandboard (OSB), and waferboard; wood panel products are less susceptible to shrinking or swelling and require less labor to install **(Figure W-7)**

Wood Post-and-Beam Framing a framework of vertical posts and horizontal beams to carry both floor and roof loads; a roof system may be conventional wood rafters or plank-and-beam framing, and a floor system may comprise conventional joists or may be plank-and-beam framing

Wood Shingle a tapered roofing shingle commonly sawn from cedar and used to cover a roof; wooden shingles may cause rapid spread of fire

Wood Structural Panel plywood, oriented strand board, and waferboard panel products that meet the requirements of UBC standard 23-2 or 23-3

Figure W-7 Wood Panel Products. *(Courtesy APA – The Engineered Wood Association.)*

Wood-Trussed Rafters a framework of wood framing members that form a light, strong, and rigid structural unit capable of relatively large clear spans; truss members are usually 2 × 4 or 2 × 6, connected by plywood gusset plates, and glued and nailed; metal gang-nail plates (gusset plates) and metal split-ring connectors (for heavy loads) are also used; typically they are engineered and fabricated and may be supported by girders or walls; examples include flat trusses, with the top or bottom chord supported 16 to 48 feet with a depth of 16 to 48 inches, and pitched trusses, like Fink and Howe trusses, with spans from 20 to 60 feet with spacing of 24 to 48 inches; NFPA 5000 requires that metal plate–connected wood trusses be manufactured in accordance with ANSI/TPI 1, "National Design Standard for Metal Plate–Connected Wood Truss Construction," which states that an approved agency shall make nonscheduled inspections of truss manufacturing and delivery operations, and that the inspection shall cover all phases of truss operations, including lumber storage, handling, cutting fixtures, presses or rollers, manufacturing, bundling, and banding **(Figures W-8a and b)**

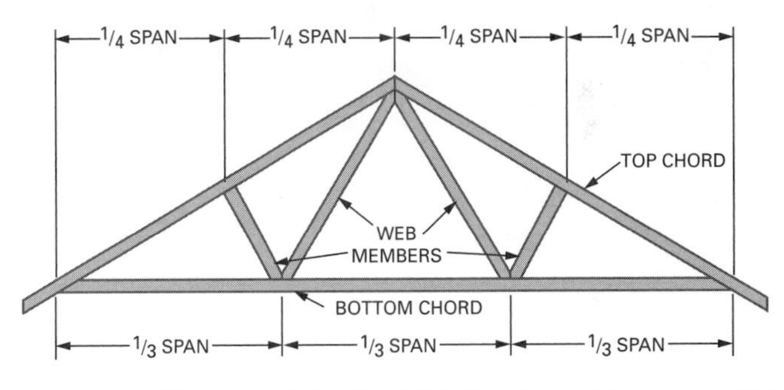

Figure W-8a Wood Trussed Rafters.

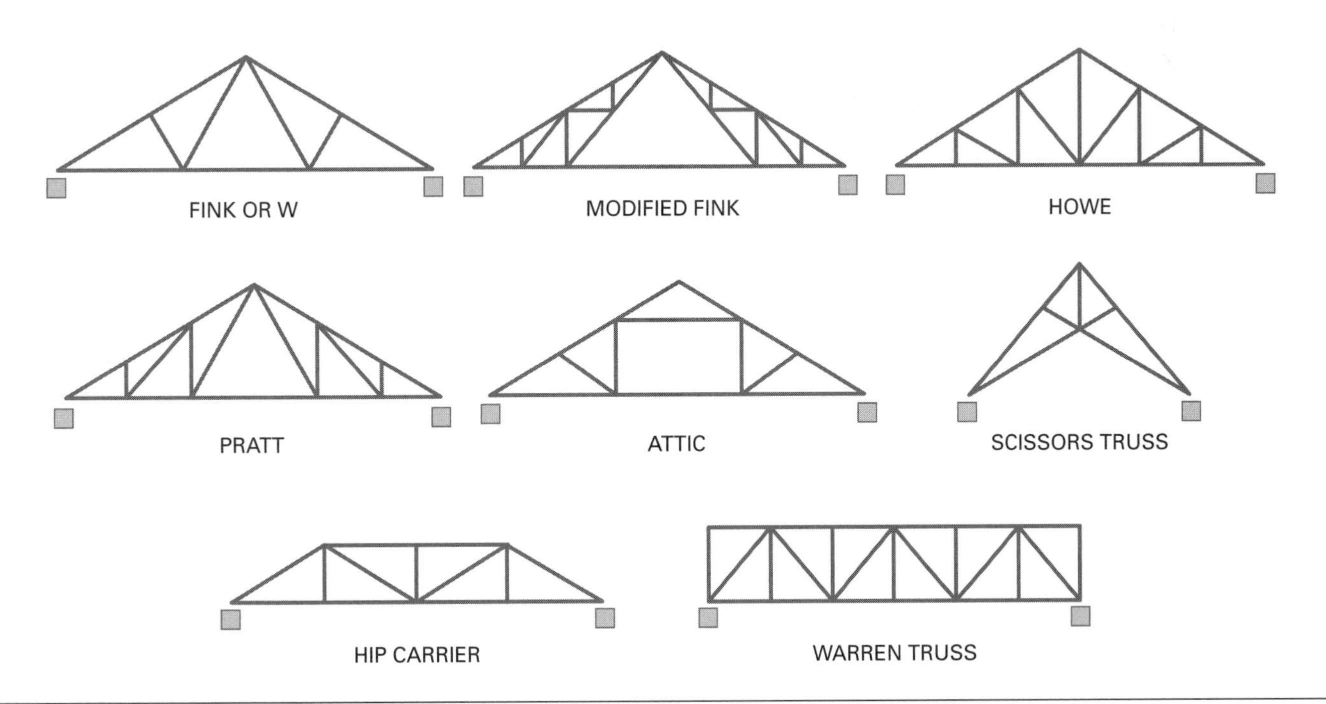

Figure W-8b Wood Trussed Rafters.

Wooden I-Beam composite wood joists with top and bottom flanges laminated and a 3/8-inch stem of plywood or OSB; the chords are slotted, and the stem is glued in place; firefighting challenges with wooden I-beam construction include plywood delamination under fire conditions, which precipitates earlier failure than with a solid beam **(Figure W-9)**

Figure W-9 Wooden I-Beam.

Working Load the total weights and stresses imposed on a structure during normal use

Woven Hip a roof covering in which the shingles overlap at the top of the hip

Woven Valley a roof covering in which the shingle courses are overlapped and then interwoven at the roof valley by overlapping alternating courses in opposite directions

⚠ **Wracking** when a component, such as a wall, is forced out of plumb

W–Type Roof Truss a type of roof truss commonly used in residential construction in which the web members form a *W;* also known as a *Fink truss* (see also *Trusses*)

Wythe single vertical row, or multiple rows, of masonry units in a wall, usually brick

X-Brace a pair of diagonal cross braces used to provide stability against wracking or sideward forces

Yard Lumber standard lumber less than 5 inches thick used in general frame construction

Yield Point the point at which strains increase without a corresponding increase in stress

Yield Strength the specified minimum yield point at which deformation occurs, measured in psi

Z-Channel Z-shaped flashing placed along a horizontal edge between abutting vertical exterior wall panels to ensure that any water that infiltrates the joint is directed outward by gravity to the outside of the wall

Zero Clearance used to describe well-insulated heating units or metal wood-burning fireplace units that do not require a specified distance between the unit and combustible surfaces

Z-Tie a metal tie used to bond masonry wall together

Additional Resources

Brannigan, V. and G. Corbett. *Brannigan's Building Construction for the Fire Service,* 4th edition, NFPA, Jones and Bartlett, Sudbury, MA, 2008.

Brunicini, A. *Fire Command,* Quincy, MA: National Fire Protection Association, 1985. Also published by Heritage Publishers, Phoenix, AZ, 2002.

California Building Code, 2007.

California Fire Code, 2007.

Ching, F. D. K. *Building Construction Illustrated,* 3rd edition. Van Nostrand-Reinhold, New York, 2001.

Dodson, D. *Fire Department Safety Officer,* 2nd edition. Thomson Delmar Learning, Clifton Park, NY, 2007.

Dunn, V. *Collapse of Burning Buildings: A Guide to Fireground Safety.* Saddle Brook, NJ: Fire Engineering Books and Videos/Penn Well, 1988.

Federal Emergency Management Agency, *Rapid Visual Screening of Buildings for Potential Seismic Hazards.* Applied Technology Council (ATC-21, ATC 154), March, 2002. Reprinted April, 2007.

Feld, J. and K. Carper. *Construction Failure,* 2nd edition. New York, NY, Wiley, 1997.

Frane, J. *Craftsman's Illustrated Dictionary of Construction Terms.* Carlsbad, CA, 1994.

Fried, E. *Fireground Tactics,* H. Chicago: Marvin Ginn Corporation, 1972.

International Fire Service Training Association, *Building Construction Related to the Fire Service,* 2nd edition, Oklahoma State University, Stillwater, OK, 1999.

International Fire Service Training Association, *Fire Service Ventilation,* 7th edition, Oklahoma State University, Stillwater, OK, 1994.

Kimbell, W. *Fire Attack 1,* Boston: National Fire Protection Association, 1966.

Layman, L. *Attacking and Extinguishing Interior Fires,* National Fire Protection Association, Boston, MA.

Mittendorf, J. *Ventilation Methods and Techniques,* Fire Technology Services, 1998.

National Fire Protection Handbook, 20th edition, Quincy, MA: National Fire Protection Association, 2008.

NFPA 1500: Standard on Fire Department Occupational Safety and Health Program, Quincy, MA: National Fire Protection Association, 2007.

NFPA 5000: Building Construction and Safety Code, Boston: National Fire Protection Association, 2009.

NFPA 251: Standard Methods of Tests of Fire Endurance of Building Construction and Materials, Boston, MA: National Fire Protection Association, 2006.

NFPA 220: Standard on Types of Building Construction, National Fire Protection Association, Boston, 2009.

NFPA 255: Standard Method of Test of Surface Burning Characteristics of Building Materials, National Fire Protection Association, Boston, 2006.

NFPA 1620: Recommended Practice for Pre-Incident Planning Boston: National Fire Protection Association, 2003.

NFPA 1670: Standard on Operations and Training for Technical Search and Rescue Incidents, Boston: National Fire Protection Association, 2009.

National Institute for Occupational Safety and Health, Firefighter Fatality Investigation and Prevention Program, available online at http://www.cdc.gov/niosh/docs/2009-100.

NIOSH Fatality Assessment and Control Evaluation (FACE) program, available online at http://www.cdc.gov/niosh/face.

"NIOSH Alert: Preventing Injuries and Death of Firefighters due to Truss System Failures," DHHS (NIOSH) Publication No. 2005-132, available online at http://www.cdc.gov/niosh/docs/2005-132.

"NIOSH Alert: Preventing Death and Injuries to Firefighters working above Fire-Damaged Floors," DHHS (NIOSH) Publication No. 2009-14, February, 2009.

"Preventing Injuries and Deaths of Firefighters," DHHS (NIOSH) Publication No. 94-125, September 1994.

"Preventing Injuries and Deaths of Fire Fighters due to Structural Collapse," DHHS (NIOSH) Publication No. 99-146, Morgantown, WV: Division of Safety Research, August 1999.

Schwartz, M. *Basic Engineering for Builders,* Craftsman Book Company, Carlsbad, CA, 1993.

Smith, M. *Building Construction: Methods and Materials for the Fire Service,* Pearson Prentice Hall, Upper Saddle River, NJ, 2008.

Uniform Building Code. International Conference of Building Officials, ICBO, Whittier, CA, 1997.

USEFUL LINKS

http://www.ul.com/global/eng/pages/offerings/industries/buildingmaterials/fire/structural

http://www.usfa.dhs.gov/fireservice/fatalities

http://www.cdc.gov/niosh/fire

http://www.firefighterclosecalls.com

http://www.firenuggets.com/about.htm

http://www.firefighternearmiss.com

http://www.woodaware.info/fireframe/index.cfm

Truss Plate Institute, http://www.tpinst.org

American Wood Council, http://www.awc.org/technical/ewpinfo.html

Many fire service on-line magazines have building construction information.

Index